For Reference

Not to be taken from this room

PATCH GUIDE

U.S. Navy Ships and Submarines

TURNER PUBLISHING COMPANY
Paducah, Kentucky

AUTHOR

The author was born in Enid, Oklahoma, enlisted in the U.S. Navy in 1957, and retired in 1979 as a Chief Aviation Machinists Mate. Since retirement he has remained in aerospace and is an executive for a major aircraft manufacturer.

Mr. Roberts is married to the former Judy White and resides in Long Beach, California. They have two children, a son Jay R. and a daughter Sherie Ann.

TURNER PUBLISHING COMPANY
The Front Line of Military History Books
412 Broadway, P. O. Box 3101
Paducah, Kentucky 42002-3101
Phone (502) 443-0121

Editor: Michael L. Roberts

Copyright © Turner Publishing Company. All rights reserved.

This book or any part thereof may not be reproduced without the written consent of the publisher.

Library of Congress
Catalog Card No.: 91-67691
ISBN: 1-56311-083-0

First Printing 1992

PREFACE

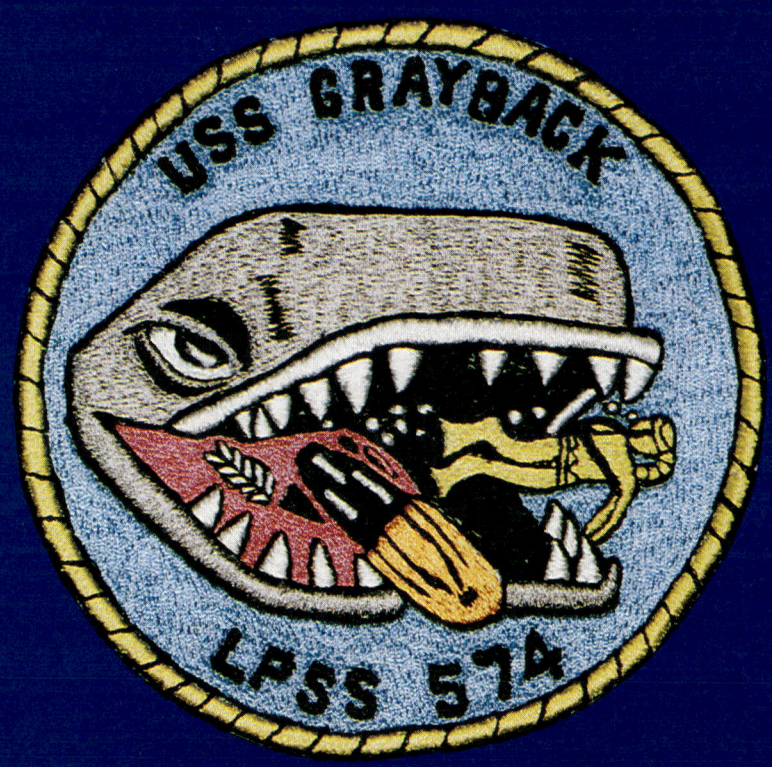

The ships crest is a customized emblem designed specifically for a particular unit. It is often displayed on plaques, patches, decals and other memorabilia. The crest builds pride in the crew and is a daily reminder that the crew belongs to the best ship in the U.S. Navy. In past years crests were designed by and for the crew members.

Although not approved by Navy regulations until the 1950s, ships have decorated their bridges and smokestacks with their crests since World War II.

Today, ship crests are authorized for "unofficial use only." Each element of the design has a specific meaning. Most crests are designed by professionals, with the design being approved by the Naval Board of Heraldry.

TABLE OF CONTENTS

Type of Ship	Designation	Page
Destroyer Tender	AD	1
Ammunition Ship	AE	3
Store Ship	AF	5
Large Auxiliary Floating Dry Dock	AFDB	6
Medium Auxiliary Floating Dry Dock	AFDM	6
Combat Store Ship	AFS	6
Miscellaneous Ship	AG	7
Escort Research Ship	AGDE	7
Auxiliary Deep Submergence Support Ship	AGDS	7
Hydrofoil Research Ship	AGEH	8
Environmental Research Ship	AGER	8
Miscellaneous Command Ship	AGF	8
Frigate Research Ship	AGFF	8
Major Communications Relay Ship	AGMR	8
Patrol Craft Tender	AGP	8
Radar Picket Ship	AGR	9
Technical Research Ship	AGTR	9
Hospital Ship	AH	9
Attack Cargo Ship	AKA	9
General Stores Issue Ship	AKS	9
Net Laying Ship	AN	9
Oiler	AO	10
Fast Combat Support Ship	AOE	12
Replenishment Oiler	AOR	13
Attack Transport Ship	APA	14
High Speed Transport Ship	APD	15
Repair Ship	AR	15
Cable Repairing Ship	ARC	16
Auxiliary Floating Dry Dock	ARD	16
Auxiliary Floating Dry Dock Medium	ARDM	17
Landing Craft Repair Ship	ARL	17
Salvage Ship	ARS	17
Submarine Tender	AS	18
Submarine Rescue Vessel	ASR	21
Auxiliary Ocean Tug	ATA	22
Fleet Ocean Tug	ATF	22
Salvage and Rescue Ship	ATS	23
Advanced Aviation Base Ship	AVB	24
Guided Missile Ship	AVM	24
Auxiliary Aircraft Landing Training Ship	AVT	24
Battleship	BB	24
Heavy Cruiser	CA	26
Guided Missile Heavy Cruiser	CAG	26
Command Ship	CC	26
Guided Missile Cruiser	CG	26
Guided Missile Cruiser (Nuclear)	CGN	32
Light Cruiser	CL	33
Guided Missile Light Cruiser	CLG	34
Aircraft Carrier	CV	34
Attack Aircraft Carrier	CVA	41
Attack Aircraft Carrier (Nuclear)	CVAN	43
Aircraft Carrier Escort	CVE	44
Aircraft Carrier Light	CVL	45
Aircraft Carrier (Nuclear)	CVN	45
Anti-Submarine Warfare Support Aircraft Carrier	CVS	48
Aircraft Carrier Training	CVT	50
Destroyer	DD	50
Escort Destroyer	DDE	61
Guided Missile Destroyer	DDG	68

TABLE OF CONTENTS (CONT'D)

Type of Ship	Designation	Page
Radar Picket Destroyer	DDR	74
Destroyer Escort	DE	75
Guided Missile Destroyer Escort	DEG	82
Radar Picket Destroyer Escort	DER	83
Frigate	DL	83
Guided Missile Frigate	DLG	83
Guided Missile Frigate (Nuclear)	DLGN	86
Deep Submergence Vehicle	DSV	86
Deep Submergence Rescue Vehicle	DSRV	86
Auxiliary Miscellaneous	EAG	86
Frigate	FF	86
Guided Missile Frigate	FFG	94
Unclassified Miscellaneous Ship	IX	101
Amphibious Command Ship	LCC	102
Amphibious Assault Ship	LHA	102
Amphibious Cargo Ship	LKA	102
Amphibious Transport Ship	LPA	103
Amphibious Assault Ship, Dock	LPD	104
Amphibious Assault Ship	LPH	105
Dock Landing Ship	LSD	107
Tank Landing Ship	LST	110
Mine Countermeasures Ship	MCM	114
Mine Countermeasures Support Ship	MCS	115
Minesweeper Coastal	MSC	115
Minesweeper Ocean (non-magnetic)	MSO	115
Nuclear Research Submarine	NR	120
Patrol	PA	118
Patrol Craft (Hydrofoil)	PCH	118
Sumarine Chaser	PCS	118
Patrol Gun Boat	PG	118
Patrol Gun Boat (Hydrofoil)	PGH	119
Patrol Combatant Missile (Hydrofoil)	PHM	119
Submarine	SS	120
Auxiliary Submarine	AGSS	124
Transport Submarine	APSS	123
Amphibious Submarine	LPSS	134
Guided Missile Submarine	SSG	134
Guided Missile Submarine (Nuclear)	SSGN	142
Anti-Submarine-Submarine	SSK	121
Radar Picket Submarine	SSR	123
Radar Picket Submarine (Nuclear)	SSRN	143
Fleet Ballistic Missile Submarine (Nuclear)	SSBN	135
Submarine (Nuclear)	SSN	143
Military Sealift Command; Store Ship	T-AF	160
Military Sealift Command; Combat Store Ship	T-AFS	160
Military Sealift Command; Miscellaneous Ship	T-AG	160
Military Sealift Command; Missile Range Instrumentation Ship	T-AGM	161
Military Sealift Command; Oceanographic Research Ship	T-AGOR	161
Military Sealift Command; Surveying Ship	T-AGS	161
Military Sealift Command; Hospital Ship	T-AH	161
Military Sealift Command; Cargo Ship	T-AK	161
Military Sealift Command; Vehicle Cargo Ship	T-AKR	162
Military Sealift Command; Oiler	T-AO	162
Military Sealift Command; Cable Repairing Ship	T-ARC	162
Mobile Noise Barge (sound lab) MONOB	YAG	163
Mobile Noise Barge	YAGR	163
Covered Lighter	YFRT	163
Large Harbor Tug, self propelled	YTB	163
Unknown (no designations)	???	163

v

ACKNOWLEDGEMENTS

Credit is given to those individuals who provided valuable advice and untiring assistance without whose help this book would not have been published. My most sincere appreciation is given to:

Larry Adams, Photographer and friend.
Jeanette Weidner, Computer Consultant and friend.
Wayne Weidner, Computer Analyst and friend.
My son Jay R. for letting me steal time from him.
My wife Judy for her support and love.

The patches depicted in this book are from the author's personal collection.

UL: USS Dixie AD-14

UC: USS Prairie AD-15

UR: USS Cascade AD-16

ML: USS Piedmont AD-17

C: USS Sierra AD-18

MR: USS Yosemite AD-19

LL: USS Yosemite AD-19

LC:

LR: USS Yosemite AD-19
 1974 Med. Cruise

UL: USS Yosemite AD-19 1974 Med. Cruise
UC: USS Hamul AD-20
UR: USS Arcadia AD-23
ML: USS Shenandoah AD-26
C: USS Isle Royal AD-29
MR: USS Tidewater AD-31
LL: USS Bryce Canyon AD-36
LC: USS Samuel Gompers AD-37
LR: USS Puget Sound AD-38

UL: USS Yellowstone AD-41　　UC: USS Acadia AD-42　　UR: USS Cape Cod AD-43

ML: USS Shenandoah AD-44　　C: USS Shasta AE-6　　MR: USS Suribachi AE-21

UL: USS Great Sitkin AE-17　　UC: USS Diamond Head AE-19　　UR: USS Mauna Kea AE-22

ML: USS Nitro AE-23　　C: USS Pyro AE-24　　MR: USS Haleakala AE-25

LL: USS Haleakala AE-25 Cap Patch　　LC: USS Kilauea AE-26　　LR: USS Butte AE-27

UL: USS Santa Barbara AE-28   UC: USS Mount Hood AE-29   UR: USS Virgo AE-30

ML: USS Chara AE-31   C: USS Flint AE-32   MR: USS Shasta AE-33

LL: USS Mount Baker AE-34   LC: USS Kiska AE-35   LR: USS Arcturus AF-52

UL: USS Rigel AF-58   UC: USS Vega AF-59   UR: Los Alamos AFDB-7

ML: Competent AFDM-6   C: Sustain AFDM-7   MR: Resolute AFDM-10

UL: USS Niagra Falls AFS-3　　UC: USS White Plains AFS-4　　UR: USS Concord AFS-5

ML: USS San Diego AFS-6　　C: USS San Jose AFS-7　　MR: USS Compass Island AG-153

LL: USS Assurance AG-521　　LC: USS Glover AGDE-1　　LR: USS Point Loma AGDS-2

UL: USS Plainview AGEH-1   UC: USS Banner AGER-1   UR: USS La Salle AGF-3

ML: USS La Salle AGF-3
Persian Gulf Yacht Club   C: USS Coronado AGF-11   MR: USS Coronado AGF-11

LL: USS Glover AGFF-1   LC: USS Arlington AGMR-2   LR: USS Graham County AGP-1176

UL: USS Watchman AGR-16 UC: USS Oxford AGTR-1 UR: USS Liberty AGTR-5

ML: USS Sanctuary AH-17 C: USS Oberon AKA-14 MR: USS Mathews AKA-96

LL: USS Rankin AKA-103 LC: USS Altair AKS-32 LR: USS Butternut AN-9

UL: USS Cimarron AO-22   **UC:** USS Chemung AO-30   **UR:** USS Guadalupe AO-32

ML: USS Maccaponi AO-41   **C:** USS Neches AO-47   **MR:** USS Ashtabula AO-51

UL: USS Caloosahatchie AO-98   UC: USS Canisteo AO-99   UR: USS Mispillion AO-105

ML: USS Waccamaw AO-109   C: USS Neosho AO-143   MR: USS Neosho AO-143

LL: USS Mississinewa AO-144   LC: USS Hassayampa AO-145   LR: USS Kawishiwi AO-146

UL: USS Truckee AO-147 UC: USS Ponchatoula AO-148 UR: USS Ponchatoula AO-148 Power and Light

ML: USS Cimarron AO-177 C: USS Monongahela AO-178 MR: USS Merrimack AO-179

UL: USS Roanoke AOR-7　　UC: USS Hunter Liggett APA-14　　UR: USS American Legion APA-17

ML: USS George Clymer APA-27　　C: USS Cavalier APA-37　　MR: USS Hansford APA-106

LL: USS Magoffin APA-199　　LC: USS Telfair APA-210　　LR: USS Navarro APA-215

UL: USS Noble APA-218 UC: USS Pickaway APA-222 UR: USS Cook APD-130

UL: USS Amphion AR-13　　UC: USS Cadmus AR-14　　UR: USS Markab AR-23

ML: USS Grand Canyon AR-28　　C: USS Neptune ARC-2　　MR: USS Aeolus ARC-3

LL: USS Waterford ARD-5　　LC: USS White Sands ARD-20　　LR: USS San Onofre ARD-30

UL: USS Oak Ridge ARDM-1
UC: USS Alamogordo ARDM-2
UR: USS Shippingport ARDM-4
ML: USS Arco ARDM-5
C: USS Sphinx ARL-24
MR: USS Escape ARS-6 Cap Patch
LL: USS Preserver ARS-8
LC: USS Safeguard ARS-25
LR: USS Bolster ARS-38

UL: USS Hoist ARS-40 UC: USS Opportune ARS-41 UR: USS Reclaimer ARS-42

ML: USS Safeguard ARS-50 C: USS Grasp ARS-51 MR: USS Salvor ARS-52

LL: USS Grapple ARS-53 LC: USS Fulton AS-11 LR: USS Sperry AS-12

UL: USS Bushnell AS-15 UC: USS Howard W. Gilmore AS-16 UR: USS Nereus AS-17

ML: USS Orion AS-18 C: USS Proteous AS-19 MR: USS Proteous AS-19

LL: USS Hunley AS-31 LC: USS Holland AS-32 LR: USS Holland AS-32

UL: USS Holland AS-32

UC: USS Simon Lake AS-33

UR: USS L.Y. Spear AS-36

ML: USS Canopus AS-34

C: USS Dixon AS-37

MR: USS Emory S. Land AS-39

LL: USS Frank Cable AS-40

LC: USS McKee AS-41

LR: USS McKee AS-41
W3 Division

UL: USS Pigeon ASR-21   UC: USS Ortolan ASR-22   UR: USS Tillamook ATA-192

ML: USS Catawba ATA-210   C: USS Kiowa ATF-72   MR: USS Mataco ATF-86

LL: USS Abnaki ATF-96   LC: USS Chowanoc ATF-100   LR: USS Hitchiti ATF-103

UL: USS Brunswick ATS-3 UC: USS Tallahatchie County AVB-2 UR: USS Norton Sound AVM-1

ML: USS Orca AVP-49 C: USS Lexington AVT-16 MR: USS North Carolina BB-55

LL: USS South Dakota BB-57 LC: USS Iowa BB-61 LR: USS New Jersey BB-62

UL: USS New Jersey BB-62

UC: USS New Jersey BB-62

UR: USS New Jersey BB-62, Peace-keeping Force Beirut, Lebanon

ML: USS New Jersey BB-62 1983-84 Beirut, Lebanon

C: USS New Jersey BB-62 Battle Group Romeo

MR: USS Missouri BB-63

LL: USS Missouri BB-63

LC: USS Missouri BB-63 1990 Rim Pac.

LR: USS Wisconsin BB-64

UL: USS Saint Paul CA-73

UC: USS Toledo CA-133

UR: USS Los Angeles CA-135

ML: USS Newport News CA-148

C: USS Newport News CA-148

MR: USS Boston CAG-1

LL: USS Northampton CC-1

LC: USS Wright CC-2

LR: USS Little Rock CG-4

UL: USS Oklahoma City CG-5　　**UC: USS Albany CG-10**　　**UR: USS Chicago CG-11**

ML: USS Columbus CG-12　　**C: USS Leahy CG-16**　　**MR: USS Harry E. Yarnell CG-17**

LL: USS Worden CG-18　　**LC: USS Dale CG-19**　　**LR: USS Richard K. Turner CG-20**

UL: USS Gridley CG-21   UC: USS England CG-22   UR: USS Halsey CG-23

ML: USS Reeves CG-24   C: USS Belknap CG-26   MR: USS Joseph Daniels CG-27

LL: USS Wainwright CG-28   LC: USS Jouett CG-29   LR: USS Horne CG-30

UL: USS Sterett CG-31   UC: USS William H. Standley CG-32   UR: USS William H. Standley CG-32

ML: USS Fox CG-33   C: USS Biddle CG-34   MR: USS Biddle CG-34

LL: USS Truxtun CG-35   LC: USS Ticonderoga CG-47   LR: USS Yorktown CG-48

UL: USS Vincennes CG-49

UC: USS Vincennes CG-49 1988 Gulf Games

UR: USS Valley Forge CG-50

ML: USS Thomas S. Gates CG-51

C: USS Bunker Hill CG-52

MR: USS Mobile Bay CG-53

LL: USS Antietam CG-54

LC: USS Leyte Gulf CG-55

LR: USS Leyte Gulf CG-55 1989 Med. Cruise

UL: USS Leyte Gulf CG-55
    Operation Desert Storm

UC: USS San Jacinto CG-56

UR: USS San Jacinto CG-56
    Operation Desert Storm

ML: USS Lake Champlain CG-57

C: USS Philippine Sea CG-58

MR: USS Princeton CG-59

LL: USS Princeton CG-59
    Operation Desert Storm

LC: USS Normandy CG-60

LR: USS Monterey CG-61

UL: USS Chancellorsville CG-62   UC: USS Cowpens CG-63   UR: USS Gettysburg CG-64

ML: USS Chosin CG-65   C: MR: USS Long Beach CGN-9

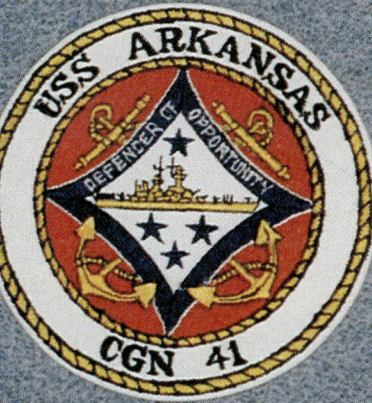

UL: USS Biloxi CL-80` UC: UR: USS Manchester CL-83

ML: USS Roanoke CL-145 C: USS Galveston CLG-3 MR: USS Little Rock CLG-4

LL: USS Springfield CLG-7 LC: USS Topeka CLG-8 LR: USS Saratoga CV-3

UL: USS Yorktown CV-5

UC: USS Wasp CV-7

UR: USS Franklin CV-13

ML: USS Ticonderoga CVS-14

C:

MR: USS Oriskany CV-34

LL: USS Midway CV-41

LC: USS Midway CV-41, 1988 Official Carrier of the Olympics

LR: USS Midway CV-41 Operation Desert Shield

UL: USS Midway CV-41
    1990-91 Persian Gulf

UC:

UR: USS Midway CV-41
    Operation Desert Storm

ML: USS Coral Sea CV-43

C:

MR: USS Midway CV-41
    Operation Desert Storm

LL: USS Forrestal CV-59

LC: USS Forrestal CV-59
    1988 North Arabian Sea

LR: USS Philippine Sea CV-47

UL: USS Forrestal CV-59
    1989 USA-USSR Summit

UC: USS Saratoga CV-60

UR: USS Saratoga CV-60
    Operation Desert Storm

ML: USS Saratoga CV-60
    Red Sea Yacht Club

C: USS Ranger CV-61

MR: USS Ranger CV-61
    C.V.S.M.

LL: USS Ranger CV-61
    1989 West. Pac.

LC:

LR: USS Ranger CV-61
    Operation Desert Shield

UL: USS Ranger CV-61
    Operation Desert Storm

UC:

UR: USS Independence CV-62

ML: USS Independence CV-62
    Med. Cruise

C:

MR: USS Independence CV-62
    1990 West. Pac.

LL: USS Independence CV-62
    1990 Iraq Pac.

LC: USS Independence CV-62
    Operation Desert Shield

LR: USS Independence CV-62
    Foreign Legion

UL: USS Kitty Hawk CV-63

UC:

UR: USS Kitty Hawk CV-63
1981 West. Pac.

ML: USS Constellation CV-64

C:

MR: USS Constellation CV-64
1987 Indian Ocean

LL: USS Constellation CV-64
1990 Around the Horn

LC: USS Constellation CV-64
Connie Does S. America

LR: USS America CV-66

UL: USS America CV-66

UC:

UR: USS America CV-66
   1979 Med. Cruise

ML: USS America CV-66
   1982 St. Thomas

C: USS America CV-66

MR: USS America CV-66
   Operation Desert Shield

LL: USS America CV-66
   Operation Desert Storm

LC: USS John F. Kennedy CV-67

LR: USS John F. Kennedy CV-67
   Drug Interdiction Cruise

UL: USS John F. Kennedy CV-67 1990 Drug Cruise

UC: USS John F. Kennedy CV-67 Operation Desert Storm

UR: USS John F. Kennedy CV-67 Operation Desert Storm

ML: USS John F. Kennedy CV-67 Operation Desert Storm

C: USS John F. Kennedy CV-67 Operation Desert Storm

MR: USS Essex CVA-9 1954-55 Far East

LL: USS Randolph CVA-15

LC:

LR: USS Hancock CVA-19

UL: USS Hancock CVA-19 UC: UR: USS Boxer CVA-21

ML: USS Bon Homme Richard CVA-31 C: USS Oriskany CVA-34 MR: USS Oriskany CVA-34 1974 West. Pac.

LL: LC: USS Shangri-La CVA-38 LR:

UL: USS Franklin D. Roosevelt CVA-42

UC: USS Coral Sea CVA-43

UR: USS Saratoga CVA-60
 1958 Med. Cruise

ML: USS Ranger CVA-61

C: USS Constellation CVA-64

MR:

LL: USS America CVA-66
 1972-73 West. Pac.

LC:

LR: USS Enterprise CVAN-65

UL: USS America CVA-66
    1974 Med. Cruise

UC:

UR: USS John F. Kennedy CVA-67

ML: USS John F. Kennedy CVA-67
    Med. Cruise

C: USS Nimitz CVAN-68

MR: USS Leyte CVE-32

LL: USS Gilbert Islands CVE-107

LC: USS Badoeing Straight CVE-116

LR: USS Palau CVE-122

UL: USS Cabot CVL-28

UC: USS San Jacinto CVL-30

UR: USS Enterprise CVN-65

ML: USS Enterprise CVN-65
1976-77 West. Pac.

C: USS Enterprise CVN-65
1983 Shellback

MR: USS Enterprise CVN-65
1989-90 World Cruise

LL: USS Enterprise CVN-65
1989-90 Just Cruising

LC: USS Nimitz CVN-68

LR: USS Nimitz CVN-68
Olympic Games

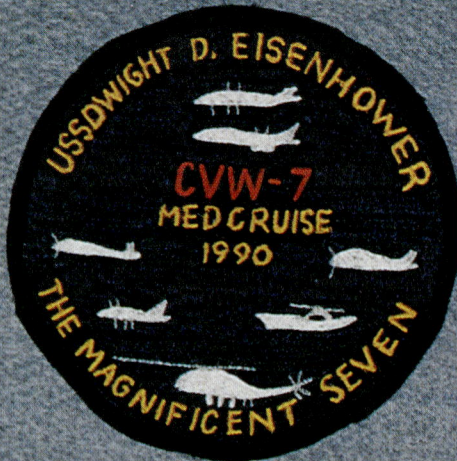

UL: USS Dwight David Eisenhower CVN-69

UC: USS Dwight David Eisenhower CVN-69, 1980 Indian Ocean

UR: USS Dwight David Eisenhower CVN-69, 1990 Med. Cruise

ML: USS Dwight David Eisenhower CVN-69, Red Sea Yacht Club

C: USS Carl Vinson CVN-70 1982 Commissioning

MR: USS Carl Vinson CVN-70

LL: USS Carl Vinson CVN-70 1983 Australia

LC: USS Carl Vinson CVN-70 1984 Admiral Flatly Award

LR: USS Carl Vinson CVN-70 1990 West. Pac.

UL: USS Carl Vinson CVN-70
     Team Spirit

UC:

UR: USS Theodore Roosevelt CVN-71

ML: USS Theodore Roosevelt CVN-71
    Operation Desert Storm

C:

MR: USS Theodore Roosevelt CVN-71
    1991 Middle East Cruise

LL: USS Theodore Roosevelt CVN-71
    Operation Desert Storm

LC:

LR: USS Theodore Roosevelt CVN-71
    Operation Desert Storm

UL: USS Theodore Roosevelt CVN-71
Operation Desert Storm

UC: USS Abraham Lincoln CVN-72

UR: USS Abraham Lincoln CVN-72
Maiden Voyage Around the Horn

ML: USS Abraham Lincoln CVN-72
Desert Storm Victory Cruise

C: USS Essex CVS-9

MR: USS Yorktown CVS-10

LL: USS Intrepid CVS-11

LC: USS Hornet CVS-12

LR: USS Hornet CVS-12

UL: USS Randolph CVS-15   UC: USS Lexington CVS-16   UR: USS Wasp CVS-18

ML: USS Bennington CVS-20   C: USS Kearsarge CVS-33   MR: USS Antietam CVS-36

LL: USS Princeton CVS-37   LC: USS Shangri-La CVS-38   LR: USS Lake Champlain CVS-39

UL: USS Lexington CVT-16

UC: USS Lexington CVT-16
1962-72 300,000 Landings

UR: CVN-65, CVN-70, CV-64
Nor. Pac. Glasnost

ML: USS Phelps DD-360

C:

MR: USS Rowan DD-405

LL: USS Chevalier DD-451

LC: USS Waller DD-466

LR: USS The Sullivans DD-537

UL: USS Twining DD-540　　　UC: USS Cowell DD-547　　　UR: USS Prichett DD-561

UL: USS Hale DD-642UC: USS Ingersoll DD-652UR: USS Bearss DD-654

ML: USS Chauncey DD-667C:MR: USS Dortch DD-670

LL: USS Picking DD-685LC: USS Remey DD-688LR: USS Allen M. Sumner DD-692

UL: USS Ingraham DD-694   UC: USS Chas. S. Sperry DD-697   UR: USS Ault DD-698

ML: USS Waldron DD-699   C: USS Hank DD-702   MR: USS Borie DD-704

LL: USS Gainard DD-706   LC: USS Wm. R. Rush DD-714   LR: USS Wiltsie DD-716

UL: USS Theodore E. Chandler DD-717   UC: USS Hamner DD-718   UR: USS Barton DD-722

ML: USS De Haven DD-727   C: USS Lyman K. Swenson DD-729   MR: USS Collett DD-730

LL: USS Blue DD-744   LC: USS Southerland DD-745   LR: USS Samuel N. Moore DD-747

UL: USS Harry E. Hubbard DD-748 UC: USS Frank E. Evans DD-754 UR: USS Strong DD-758

ML: USS Lofberg DD-759 C: USS John W. Thomason DD-760 MR: USS Henley DD-762

L: USS William C. Lawe DD-763 LC: USS Lloyd Thomas DD-764 LR: USS Keppler DD-765

UL: USS Zellars DD-777

UC: USS R. K. Huntington DD-781

UR: USS Gurke DD-783

ML: USS McKean DD-784

C: USS Henderson DD-785

MR: USS Richard B. Anderson DD-786

LL: USS James E. Kyes DD-787

LC: USS Hollister DD-788

LR: USS Eversole DD-789

UL: USS Rich DD-820

UC: USS Johnston DD-821

UR: USS Johnston DD-821

ML: USS R.H. McCard DD-822

C: USS Samuel B. Roberts DD-823

MR: USS Carpenter DD-825

LL: USS Robert A. Owens DD-827

LC: USS Herbert J. Thomas DD-833

LR: USS Charles P. Cecil DD-835

UL: USS George K. Mackenzie DD-836   UC: USS George K. Mackenzie DD-836   UR: USS Power DD-839

ML: USS Glennon DD-840   C: USS Noa DD-841   MR: USS Fiske DD-842

LL: USS Warrington DD-843   LC: USS Bausell DD-845   LR: USS Ozbourn DD-846

UL: USS Robert Wilson DD-847   UC: USS R.E. Kraus DD-849   UR: USS Rupertus DD-851

ML: USS Charles H. Roan DD-853   C: USS Fred T. Berry DD-858   MR: USS McCaffery DD-860

LL: USS Vogelgesang DD-862   LC: USS Steinaker DD-863   LR: USS Harold J. Ellison DD-864

UL: USS Cone DD-866　　UC: USS Stribling DD-867　　UR: USS Brownson DD-868

ML: USS Arnold J. Isbell DD-869　　C: USS Damato DD-871　　MR: USS Hawkins DD-873

LL: USS Henry W. Tucker DD-875　　LC: USS Rogers DD-876　　LR: USS Dyess DD-880

UL: USS Furse DD-882   UC: USS N.K. Perry DD-883   UR: USS Floyd B. Parks DD-884

ML: USS Orleck DD-886   C: USS Meredith DD-890   MR: USS Forest Sherman DD-931

UL: USS Manley DD-940UC: USS Dupont DD-941UR: USS Bigelow DD-942

ML: USS Blandy DD-943C: USS Mullinix DD-944MR: USS Hull DD-945

LL: USS Edson DD-946LC: USS Morton DD-948LR: USS Richard S. Edwards DD-950

UL: USS Turner Joy DD-951　　UC: USS Spraunce DD-963　　UR: USS Paul F. Foster DD-964

ML: USS Kinkaid DD-965　　C: USS Hewitt DD-966　　MR: USS Elliot DD-967

LL: USS Arthur W. Radford DD-968　　LC: USS Peterson DD-969　　LR: USS Caron DD-970

UL: USS David R. Ray DD-971   UC: USS Oldendorf DD-972   UR: USS John Young DD-973

ML: USS Comte De Grasse DD-974   C: USS Comte De Grasse DD-974 1983 Stanavforlant   MR: USS O'Brien DD-975

L: USS Merrill DD-978   LC: USS Briscoe DD-977   LR: USS Stump DD-978

UL: USS Conolly DD-979   UC: USS Moosbrugger DD-980   UR: USS John Hancock DD-981

ML: USS Nicholson DD-982   C: USS John Rodgers DD-983   MR: USS Leftwich DD-984

LL: USS Cushing DD-985   LC: USS Harry W. Hill DD-986   LR: USS O'Bannon DD-987

UL: USS Thorn DD-988   UC: USS Deyo DD-989   UR: USS Ingersoll DD-990

ML: USS Fife DD-991   C: USS Fletcher DD-992   MR: USS Hayler DD-997

UL: USS Charles F. Adams DDG-2   UC: USS John King DDG-3   UR: USS Lawrence DDG-4

ML: USS Claude V. Ricketts DDG-5   C: USS Barney DDG-6   MR: USS Henry B. Wilson DDG-7

LL: USS Lynde McCormick DDG-8   LC: USS Towers DDG-9   LR: USS Sampson DDG-10

UL: USS Sampson DDG-10
    Operation Desert Shield

UC: USS Sellers DDG-11

UR: USS Robison DDG-12

ML: USS Hoel DDG-13

C: USS Buchanan DDG-14

MR: USS Berkley DDG-15

LL: USS Joseph Strauss DDG-16

LC: USS Conyngham DDG-17

LR: USS Semmes DDG-18

UL: USS Tattnall DDG-19   UC: USS Goldsborough DDG 20   UR: USS Cockrane DDG-21

ML: USS Benjamin Stoddert DDG-22   C: MR: USS Richard E. Byrd DDG-23

LL: USS Waddell DDG-24   LC: LR: USS Waddell DDG-24

UL: USS Farragut DDG-37   UC: USS Luce DDG-38   UR: USS MacDonough DDG-39

UL: USS Dahlgren DDG-43
1991 Round The Horn

UC: USS William V. Pratt DDG-44

UR: USS Dewey DDG-45

ML: USS Preble DDG-46

C: USS Arleigh Burke DDG-51

MR: USS Kidd DDG-993

LL: USS Callaghan DDG-994

LC: USS Scott DDG-995

LR: USS Chandler DDG-996

UL: USS Eugene A. Greene DDR-711   UC: USS W.R. Rush DDR-714   UR: USS William M. Wood DDR-715

ML: USS Myles C. Fox DDR-829   C: USS Goodrich DDR-831   MR: USS Charles P. Cecil DDR-835

LL: USS Ernest G. Small DDR-838   LC: USS Vesole DDR-878   LR: USS Leary DDR-879

UL: USS Bridget DE-1024   UC: USS Bauer DE-1025   UR: USS Hooper DE-1026

ML: USS Joseph K. Taussig DE-1030   C: USS John R. Perry DE-1034   MR: USS McMorris DE-1036

LL: USS Bronstein DE-1037   LC: USS Edward McDonnell DE-1043   LR: USS Davidson DE-1045

UL: USS Voge DE-1047

UC: USS Sample DE-1048

UR: USS Koelsch DE-1049

ML: USS Albert David DE-1050

C: USS O'Callahan DE-1051

MR: USS Knox DE-1052

LL: USS Roark DE-1053

LC: USS Gray DE-1054

LR: USS Connole DE-1056

UL: USS W.S. Sims DE-1059   **UC:** USS Lang DE-1060   **UR:** USS Patterson DE-1061

ML: USS Whipple DE-1062   **C:** USS Reasoner DE-1063   **MR:** USS Lockwood DE-1064

LL: USS Stein DE-1065   **LC:** USS Marvin Shields DE-1066   **LR:** USS Francis Hammond DE-1067

UL: USS Vreeland DE-1068   **UC:** USS Bagley DE-1069   **UR:** USS Downes DE-1070

ML: USS Badger DE-1071   **C:** USS Blakely DE-1072   **MR:** USS Robert E. Peary DE-1073

UL: USS Fanning DE-1076   UC: USS Joseph Hewes DE-1078   UR: USS Bowen DE-1079

ML: C: MR:

UL: USS Elmer Montgomery DE-1082   UC: USS Cook DE-1083   UR: USS Donald B. Beary DE-1085

ML: C: MR:

LL: USS Brewton DE-1086   LC: USS Kirk DE-1087   LR: USS Jesse L. Brown DE-1089

UL: USS Thomas C. Hart DE-1092   UC: USS Truett DE-1095   UR: USS Valdez DE-1096

ML: USS Moinester DE-1097   C: USS Brooke DEG-1   MR: USS Ramsey DEG-2

LL: USS Schofield DEG-3   LC: USS Talbot DEG-4   LR: USS Richard L. Page DEG-5

UL: USS Falgout DER-324 **UC:** USS Brister DER-327 **UR:** USS Kretchmer DER-329

ML: USS Forster DER-334 **C:** USS Savage DER-386 **MR:** USS Haverfield DER-393

UL: USS MacDonough DLG-8   UC: USS King DLG-10   UR: USS Mahan DLG-11

ML: USS William V. Pratt DLG-13   C: USS Dewey DLG-14   MR: USS Preble DLG-15

LL: USS Harry E. Yarnell DLG-17   LC: USS England DLG-22   LR: USS Reeves DLG-24

UL: USS Belknap DLG-26 UC: USS Josephus Daniels DLG-27 UR: USS Wainwright DLG-28

ML: USS Jouett DLG-29 C: USS Horne DLG-30 MR: USS Wm. H. Standley DLG-32

UL: USS Bainbridge DLGN-25

UC: USS Truxton DLGN-35

UR: Trieste II DSV-1

ML: Turtle DSV-3

C: Sea Cliff DSV-4

MR: Mystic DSRV-1

LL: Avalon DSRV-2

LC: USS Observation Island EAG-154

LR: USS Julius A. Furer FF-6

UL: USS Bronstein FF-1037   UC: USS McCloy FF-1038   UR: USS Garcia FF-1040

L: USS Bradley FF-1041   C: USS McDonnell FF-1043   MR: USS Brumby FF-1044

L: USS Davidson FF-1045   LC: USS Voge FF-1047   LR: USS Sample FF-1048

UL: USS Meyerkord FF-1058 UC: USS W.S. Sims FF-1059 UR: USS Lang FF-1060

ML: USS Patterson FF-1061 C: USS Whipple FF-1062 MR: USS Reasoner FF-1063

UL: USS Marvin Shields FF-1066 UC: USS Francis Hammond FF-1067 UR: USS Vreeland FF-1068

ML: USS Bagley FF-1069 C: USS Downes FF-1070 MR: USS Badger FF-1071

LL: USS Blakely FF-1072 LC: USS Robert E. Peary FF-1073 LR: USS Harold E. Holt FF-1074

UL: USS Trippe FF-1075   UC: USS Fanning FF-1076   UR: USS Ouellet FF-1077

ML: USS Joseph Hewes FF-1078   C: USS Bowen FF-1079   MR: USS Paul FF-1080

LL: USS Aylwin FF-1081   LC: LR: USS Elmer Montgomery FF-1082

UL: USS Cook FF-1083   **UC:** USS McCandless FF-1084   **UR:** USS Donald B. Beary FF-1085

ML: USS Brewton FF-1086   **C:** USS Kirk FF-1087   **MR:**

LL: USS Jessie L. Brown FF-1089   **LC:** USS Ainsworth FF-1090   **LR:** USS Barbey FF-1088

UL: USS Miller FF-1091   UC: USS Thomas C. Hart FF-1092   UR: USS Capodanno FF-1093

ML: USS Pharris FF-1094   C: USS Truett FF-1095   MR: USS Valdez FF-1096

UL: USS Brooke FFG-1   UC: USS Ramsey FFG-2   UR: USS Schofield FFG-3

ML: USS Talbot FFG-4   C: USS Richard L. Page FFG-5   MR: USS Julias A. Furer FFG-6

LL: USS Oliver Hazard Perry FFG-7   LC: USS McInerney FFG-8   LR: USS Wadsworth FFG-9

UL: USS Wadsworth FFG-9

UC: USS Wadsworth FFG-9

UR: USS Duncan FFG-10

ML:

C:

MR:

LL: USS Clark FFG-11

LC: USS George Philip FFG-12

LR: USS Samuel Eliot Morison FFG-13

UL: USS Sides FFG-14   UC: USS Estocin FFG-15   UR: USS Clifton Sprague FFG-16

ML: USS John A. Moore FFG-19   C: USS Antrim FFG-20   MR: USS Flatley FFG-21

LL: USS Fahrion FFG-22   LC: USS Lewis B. Puller FFG-23   LR: USS Jack Williams FFG-24

UL: USS Copeland FFG-25 UC: USS Gallery FFG-26 UR: USS Mahlon S. Tisdale FFG-27

ML: USS Boone FFG-28 C: USS Stephen W. Groves FFG-29 MR: USS Reid FFG-30

LL: USS Stark FFG-31 LC: USS John L. Hall FFG-32 LR: USS Jarrett FFG-33

UL: USS Aubrey Fitch FFG-34   UC: USS Underwood FFG-36   UR: USS Crommelin FFG-37

ML: USS Curts FFG-38   C: USS Doyle FFG-39   MR: USS Halyburton FFG-40

UL: USS De Wert FFG-45   UC: USS Rentz FFG-46   UR: USS Nicholas FFG-47

ML: USS Vandegrift FFG-48   C: USS Robert G. Bradley FFG-49   MR: USS Taylor FFG-50

UL: USS Ford FFG-54

UC:

UR: USS Ford FFG-54
    Operation Desert Storm

ML: USS Elrod FFG-55

C: USS Reuben James FFG-57
   Plankowner

MR: USS Simpson FFG-56

LL:

LC: USS Reuben James FFG-57
    Engineering Dept., Plankowner

LR:

UL: USS Reuben James FFG-57
UC: USS Samuel B. Roberts FFG-58
UR: USS Samuel B. Roberts FFG-58 Operation Desert Shield
ML: USS Kauffman FFG-59
C: USS Rodney M. Davis FFG-60
MR: USS Ingraham FFG-61

UL: USS Blue Ridge LCC-19　　UC: USS Mount Whitney LCC-20　　UR: USS Tarawa LHA-1

ML: USS Saipan LHA-2　　C: USS Saipan LHA-2 1985 Med. Cruise　　MR: USS Belleau Wood LHA-3

UL: USS Tulare LKA-112
UC: USS Tulare LKA-112 Cap Patch
UR: USS Charleston LKA-113

ML: USS Durham LKA-114
C: USS Mobile LKA-115
MR: USS St. Louis LKA-116

LL: USS El Paso LKA-117
LC: USS Paul Revere LPA-248
LR: USS Francis Marion LPA-249

UL: USS Raleigh LPD-1 UC: USS Vancouver LPD-2 UR: USS Austin LPD-4

UL: USS Juneau LPD-10
UC: USS Coronado LPD-11
UR: USS Shreveport LPD-12
ML: USS Nashville LPD-13
C: USS Trenton LPD-14
MR: USS Ponce LPD-15
LL: USS Iwo Jima LPH-2
LC: USS Okinawa LPH-3
LR: USS Okinawa LPH-3 Operation Desert Shield

UL: USS Guadalcanal LPH-7

UC: USS Valley Forge LPH-8

UR: USS Valley Forge LPH-8
Operation Desert Shield

ML: USS Guam LPH-9

C: USS Tripoli LPH-10

MR: USS Tripoli LPH-10
Operation Desert Shield

LL: USS New Orleans LPH-11

LC: USS Inchon LPH-12

LR: USS Inchon LPH-12
Operation Desert Storm

UL: USS Ashland LSD-1　　UC: USS Epping Forest LSD-4　　UR: USS Gunston Hall LSD-5

ML: USS Lind Enwald LSD-6　　C: USS Oak Hill LSD-7　　MR: USS Shadwell LSD-15

LL: USS Fort Mandan LSD-21　　LC: USS Fort Marion LSD-22　　LR: USS San Marcos LSD-25

UL: USS Thomaston LSD-28   UC: USS Plymouth Rock LSD-29   UR: USS Plymouth Rock LSD-29

ML: USS Fort Snelling LSD-30   C: USS Point Defiance LSD-31   MR: USS Spiegal Grove LSD-32

LL: USS Alamo LSD-33   LC: USS Hermitage LSD-34   LR: USS Monticello LSD-35

UL: USS Anchorage LSD-36   UC: USS Portland LSD-37   UR: USS Pensacola LSD-38

ML: USS Mount Vernon LSD-39   C: USS Fort Fisher LSD-40   MR: USS Whidbey Island LSD-41

LL: USS Germantown LSD-42   LC: USS Fort McHenry LSD-43   LR: USS Gunston Hall LSD-44

UL: USS Comstock LSD-45

UC: USS Duval County LST-758

UR: USS Garrett County LST-786

ML: USS Hunterdon County LST-838

C: USS Kemper County LST-854

MR: USS Litchfield County LST-901

LL: USS Luzerne County LST-902

LC: USS Page County LST-1076

LR: USS Snohomish County LST-1126

UL: USS Stark County LST-1134   UC: USS Stone County LST-1141   UR: USS Talahatchie County LST-1154

ML: USS Tioga County LST-1158   C: USS Vernon County LST-1161   MR: USS Waldo County LST-1163

LL: USS De Soto County LST-1171   LC: LR: USS Suffolk County LST-1173

UL: USS Grant County LST-1174   UC: USS York County LST-1175   UR: USS Lorain County LST-1177

ML: USS Wood County LST-1178   C: USS Newport LST-1179   MR: USS Manitowoc LST-1180

LL: USS Sumter LST-1181   LC: LR: USS Fresno LST-1182

UL: USS Peoria LST-1183 UC: USS Frederick LST-1184 UR: USS Schenectady LST-1185

ML: USS Cayuga LST-1186 C: USS Tuscaloosa LST-1187 MR: USS Saginaw LST-1188

L: USS San Bernardino LST-1189 LC: LR: USS Boulder LST-1190

UL: USS Racine LST-1191   UC: USS Spartanburg County LST-1192   UR: USS Fairfax County LST-1193

ML: USS La Moure County LST-1194   C: USS Barbour County LST-1195   MR: USS Harlan County LST-1196

LL: USS Barnstable County LST-1197   LC: USS Bristol County LST-1198   LR: USS Avenger MCM-1

UL: USS Sentry MCM-3

UC: USS Devastator MCM-6

UR: USS Catskill MCS-1

ML: USS Ozark MCS-2

C: USS Epping Forest MCS-7

MR: USS Frigate Bird MSC-191

LL: USS Vireo MSC-205

LC: USS Whippoorwill MSC-207

LR: USS Conflict MSO-426

UL: USS Constant MSO-427
UC: USS Constant MSO-427 Operation Desert Storm
UR: USS Dash MSO-428

ML: USS Detector MSO-429
C:
MR: USS Direct MSO-430

LL: USS Dominant MSO-431 Cap Patch
LC: USS Engage MSO-433
LR: USS Enhance MSO-437

UL: USS Excel MSO-439   UC: USS Exultant MSO-441   UR: USS Fearless MSO-442

ML: USS Fidelity MSO-443   C: USS Fortify MSO-446   MR: USS Illusive MSO-448

LL: USS Implicit MSO-455   LC: USS Inflict MSO-456   LR: USS Pluck MSO-464

UL: USS Salute MSO-470

UC: USS Conquest MSO-488

UR: USS Gallant MSO-489

ML: USS Pledge MSO-492

C: USS Affray MSO-511

MR: USS Cambria PA-36

LL: USS High Point PCH-1

LC: USS Hollidaysburg PCS-1385

LR: USS Gallup PG-85

UL: USS Antelope PG-86   UC: USS Crockett PG-88   UR: USS Marathon PG-89

ML: USS Tacoma PG-92   C: USS Grand Rapids PG-98   MR: USS Beacon PG-99

LL: USS Greenbay PG-101   LC: USS Flagstaff PGH-1   LR: USS Pegasus PHM-1

UL: USS Hercules PHM-2
UC: USS Taurus PHM-3
UR: USS Aquila PHM-4

ML: USS Aries PHM-5
C: USS Gemini PHM-6
MR: USS Nuclear Research Submarine NR-1

LL: USS Porpoise SS-172
LC: USS Permit SS-178
LR: USS Pollack SS-180

UL: USS Snapper SS-185   UC: UR: USS Stingray SS-186

ML: USS Sturgeon SS-187   C: USS Spearfish SS-190 Battle Flag   MR: USS Greenling SS-213

UL: USS Angler SS-240

UC:

UR: USS Bluegill SSK-242

ML: USS Bream SS-243

C: USS Cavalla SS-244

MR: USS Croaker SS-246

LL: USS Flasher SS-249

LC:

LR: USS Rasher SS-269

UL: USS Ray SSR-271   UC: UR: USS Sawfish SS-276

ML: USS Tunny SS-282   C: MR: USS Tunny APSS-282

UL: USS Batfish SS-310   UC: USS Archerfish AGSS-311 Operation Seascan   UR: USS Perch APSS-313

ML: USS Barbero SSG-317   C: MR: USS Barbero SSG-317

UL: USS Becuna SS-319UC: USS Bergall SS-320UR: USS Blackfin SS-322

ML: USS Criman SS-323C:MR: USS Bullhead SS-332

LL: USS Carbonero SS-337LC:LR: USS Carbonero SS-337
1965 West. Pac.

UL: USS Carp SS-338   UC: USS Catfish SS-339   UR: USS Entemedor SS-340

ML: USS Clamagore SS-343   C: USS Cobbler SS-344   MR: USS Corporal SS-346

UL: USS Diodon SS-349   **UC:** USS Dogfish SS-350   **UR:** USS Halfbeak SS-352

ML: USS Hardhead SS-365   **C:** USS Icefish SS-367   **MR:** USS Lamprey SS-372

LL: USS Sandlance SS-381   **LC:**   **LR:** USS Parche SS-384

UL: USS Bang SS-385   UC: USS Bang SS-385   UR: USS Piranha SS-389

ML: USS Pomfret SS-391   C: MR: USS Sterlet SS-392

UL; USS Segundo SS-398   UC: USS Sea Cat SS-399   UR: USS Sea Devil SS-400

ML: USS Sea Dog SS-401   C: USS Sea Fox SS-402   MR: USS Atule SS-403

UL: USS Argonaut SS-475   UC: USS Runner SS-476   UR: USS Conger SS-477

ML: USS Cutlass SS-478   C: MR: USS Cutlass SS-478

UL: USS Requin SSR-481   UC: USS Irex SS-482   UR: USS Odax SS-484

ML: USS Remora SS-487   C: MR: USS Sarda SS-488

LL: USS Spinax SSR-489   LC: USS Gudgeon SS-507   LR: USS Amber Jack SS-522

UL: USS Dolphin SS-555　　UC: USS Tang SS-563　　UR: USS Trigger SS-564

ML: USS Wahoo SS-565　　C: USS Trout SS-566　　MR: USS Gudgeon SS-567

LL: USS Harder SS-568　　LC: USS Sailfish SS-572　　LR: USS Salmon SS-573

UL: USS Grayback LPSS-574   UC: USS Darter SS-576   UR: USS Growler SSG-577

ML: USS Barbel SS-580   C: USS Barbel SS-580   MR: USS Blueback SS-581

L: USS George Washington SSBN-598　　UC: USS George Washington SSBN-598 1959-85, Decommissioning Crew　　UR: USS Patrick Henry SSBN-599

IL: USS Theodore Roosevelt SSBN-600　　C: USS Robert E. Lee SSBN-601　　MR: USS Abraham Lincoln SSBN-602

· USS Ethan Allen SSBN-608　　LC: USS Sam Houston SSBN-609　　LR: USS Thomas A. Edison SSBN-610

UL: USS John Marshall SSBN-611

UC: USS Lafayette SSBN-616

UR: USS Alexander Hamilton SSBN-617

ML: USS Thomas Jefferson SSBN-618

C: USS Thomas Jefferson SSBN-618 Purdums Pirates

MR: USS Thomas Jefferson SSBN-618 1963-1985, Decommissioning Crew

LL: USS Andrew Jackson SSBN-619

LC: USS John Adams SSBN-620

LR:

UL: USS James Monroe SSBN-622   UC: USS Nathan Hale SSBN-623   UR: USS Woodrow Wilson SSBN-624

ML: USS Henry Clay SSBN-625   C: USS Daniel Webster SSBN-626   MR: USS James Madison SSBN-627

LL: USS Daniel Boone SSBN-629   LC: LR: USS John C. Calhoun SSBN-630

UL: USS U.S. Grant SSBN-631 UC: UR: USS Von Steuben SSBN-632

ML: USS Casimir Pulaski SSBN-633 C: USS Stonewall Jackson SSBN-634 MR: USS Sam Rayburn SSBN-635

LL: USS Nathanael Greene SSBN-636 LC: USS Benjamin Franklin SSBN-640 LR: USS Simon Bolivar SSBN-641

UL: USS Kamehameha SSBN-642
UC: USS George Bancroft SSBN-643
UR: USS Lewis & Clark SSBN-644

ML: USS James K. Polk SSBN-645
C: USS George C. Marshall SSBN-654
MR: USS Henry L. Stimson SSBN-655

LL: USS George Washington Carver SSBN-656
LC: USS Francis Scott Key SSBN-657
LR: USS Mariano G. Vallejo SSBN-658

UL: USS Will Rogers SSBN-659
UC: USS Ohio SSBN-726
UR: USS Michigan SSBN-727

ML: USS Florida SSN-728 Launch Crew
C: USS Florida SSBN-728
MR: USS Georgia SSBN-729

LL: USS Henry M. Jackson SSBN-730
LC: USS Alabama SSBN-731 Launch Crew
LR:

UL: USS Alabama SSBN-731

UC: USS Alaska SSBN-732 Launch Crew

UR: USS Alaska SSBN-732

ML: USS Alaska SSBN-732

C: USS Nevada SSBN-733 Launch Crew

MR: USS Nevada SSBN-733

LL: USS Tennessee SSBN-734 Launch Crew

LC:

LR: USS Tennessee SSBN-734

UL: USS Pennsylvania SSBN-735   UC: UR: USS West Virginia SSBN-736

ML: USS Kentucky SSBN-737   C: USS Halibut SSGN-587   MR: USS Neptune SSN-525
                                                                         Movie "Gray Lady Down"

LL: USS Nautilus SSN-571   LC: USS Seawolf SSN-575   LR: USS Skate SSN-578

UL: USS Swordfish SSN-579 UC: USS Sargo SSN-583 UR: USS Seadragon SSN-584

ML: USS Skipjack SSN-585 C: USS Triton SSN-586 MR: USS Triton SSRN-586

LL: USS Halibut SSN-587 LC: USS Scamp SSN-588 LR: USS Scorpion SSN-589

UL: USS Sculpin SSN-590   UC: USS Shark SSN-591   UR: USS Snook SSN-592

ML: USS Snook SSN-592   C: USS Thresher SSN-593   MR: USS Permit SSN-594
    1981 Med. Run

LL: USS Permit SSN-594   LC: USS Plunger SSN-595   LR: USS Barb SSN-596
    1976 West. Pac. Patrol

UL: USS Tullibee SSN-597   UC: USS Pollack SSN-603   UR: USS Haddo SSN-604

ML: USS Jack SSN-605   C: USS Tinosa SSN-606   MR: USS Dace SSN-607

LL: USS Guardfish SSN-612   LC: USS Flasher SSN-613   LR: USS Flasher SSN-613

UL: USS Greenling SSN-614

UC:

UR: USS Gato SSN-615

ML: USS Haddock SSN-621

C: USS Tecumseh SSBN-628

MR: USS Sturgeon SSN-637

LL: USS Whale SSN-638

LC: USS Whale SSN-638 Power and Light

LR: USS Tautog SSN-639

UL: USS Grayling SSN-646 UC: USS Pogy SSN-647 UR: USS Aspro SSN-648

ML: USS Sunfish SSN-649 C: USS Pargo SSN-650 MR: USS Queenfish SSN-651

LL: USS Puffer SSN-652 LC: USS Ray SSN-653 LR: USS Sand Lance SSN-660

UL: USS Lapon SSN-661UC: USS Gurnard SSN-662UR: USS Hammerhead SSN-663

ML: USS Sea Devil SSN-664C: USS Guitarro SSN-665MR: USS Hawkbill SSN-666

LL: USS Bergall SSN-667LC: USS Spadefish SSN-668LR: USS Seahorse SSN-669

UL: USS Finback SSN-670 UC: USS Narwhal SSN-671 Blue Crew UR: USS Pintado SSN-672

ML: USS Flying Fish SSN-673 C: USS Flying Fish SSN-673 1989-90 Med. Run MR: USS Trepang SSN-674

UL: USS Archerfish SSN-678   UC: USS Silversides SSN-679   UR: USS William H. Bates SSN-680

ML: USS Batfish SSN-681   C: MR: USS Tunny SSN-682

LL: USS Parche SSN-683   LC: USS Cavalla SSN-684   LR: USS Glenard P. Lipscomb SSN-68

UL: USS L. Mendel Rivers SSN-686

UC: USS Richard B. Russell SSN-687

UR: USS Richard B. Russell SSN-687
1990 Christmas Cruise, Blue Crew

ML: USS Richard B. Russell SSN-687
1991 Turban Tour

C:

MR: USS Los Angeles SSN-688

LL: USS Los Angeles SSN-688

LC: USS Baton Rouge SSN-689

LR: USS Philadelphia SSN-690

UL: USS Memphis SSN-691UC: USS Omaha SSN-692UR: USS Cincinnati SSN-693

ML: USS Groton SSN-694C:MR: USS Birmingham SSN-695

LL: USS Birmingham SSN-695
Battle "E" & Admin. "A"LC:LR: USS New York City SSN-696

UL: USS Indianapolis SSN-697   UC: UR: USS Bremerton SSN-698

ML: USS Jacksonville SSN-699   C: MR: USS Dallas SSN-700

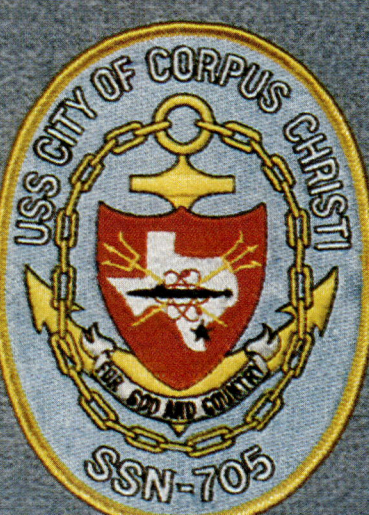

UL: USS Boston SSN-703
    Launch Crew

UC:

UR: USS Boston SSN-703

ML: USS Baltimore SSN-704
    Launch Crew

C:

MR: USS Baltimore SSN-704

LL: USS Corpus Christi SSN-705
    Launch Crew

LC: USS Corpus Christi SSN-705

LR: USS City of Corpus Christi
    SSN-705

UL: USS Albuquerque SSN-706
UC: USS Portsmouth SSN-707
UR: USS Minneapolis, St. Paul SSN-708
ML: USS Hyman G. Rickover SSN-709
MR: USS Agusta SSN-710
LL: USS San Francisco SSN-711 Launch Crew
LR: USS San Francisco SSN-711

UL: USS Atlanta SSN-712 Launch Crew

UC: USS Atlanta SSN-712

UR: USS Atlanta SSN-712 1982 Autec

ML: USS Atlanta SSN-712 1983 Deployment

C: USS Houston SSN-713 Launch Crew

MR: USS Houston SSN-713

LL: USS Norfolk SSN-714 Plankowner

LC: USS Norfolk SSN-714 Launch Crew

LR: USS Norfolk SSN-714

UL: USS Buffalo SSN-715
    Launch Crew

UC: USS Buffalo SSN-715

UR: USS Salt Lake City SSN-716
    Launch Crew

ML: USS Salt Lake City SSN-716

C: USS Olympia SSN-717

MR: USS Honolulu SSN-718

LL: USS Providence SSN-719

LC: USS Pittsburgh SSN-720

LR: USS Chicago SSN-721
    Cap Patch

UL: USS Chicago SSN-721

UC: USS Key West SSN-722

UR: USS Oklahoma City SSN-723

ML: USS Louisville SSN-724 Launch Crew

C: USS Louisville SSN-724

MR: USS Helena SSN-725 Launch Crew

LL: USS Helena SSN-725

LC: USS Newport News SSN-750

LR: USS San Juan SSN-751

UL: USS Pasadena SSN-752
    1989 Commissioning

UC:

UR: USS Pasadena SSN-752

ML: USS Albany SSN-753

C: USS Topeka SSN-754
   Pre-Commissioning Unit

MR: USS Topeka SSN-754

LL: USS Miami SSN-755

LC:

LR: USS Alexandria SSN-757

UL: USS Ashville SSN-758   UC: UR: USS Jefferson City SSN-759

UL: USNS Hayes T-AGOR-16
UC: USNS Vandenberg T-AGM-10
UR: USNS Vanguard T-AGM-19

ML: USNS Redstone T-AGM-20
C: USNS Bowditch T-AGS-21
MR: USNS Elisha Kent Kane T-AGS-27

LL: USNS Harkness T-AGS-32
LC: USNS Mercy T-AH-19
LR: USNS Mirfak T-AK-271

UL: USNS Norwalk T-AK-279
UC: USNS Marshfield T-AK-282
UR: USNS Mercury T-AKR-10
ML: USNS Jupiter T-AKR-11
C: USNS Marias T-AO-57
MR: USNS Waccamaw T-AO-109
LL: USNS Henry J. Kaiser T-A0-187
LC: USNS Neptune T-ARC-2
LR: USNS Aeolus T-ARC-3

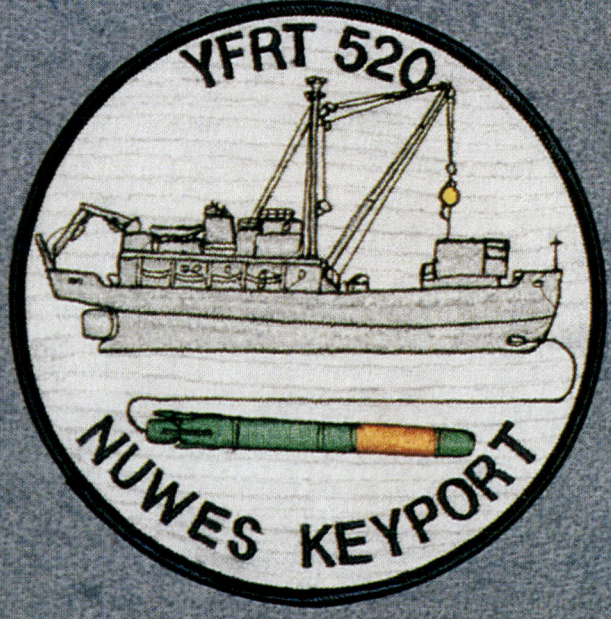

UL: USS George Eastman YAG-39

UC: USS Monob YAG-61

UR: USS Interceptor YAGR-8

ML: USS Investigator YAGR Division 21

C:

MR: USS YFRT-520 Nuwes; Keyport, WA.

LL: USS Apopka YTB-778

LC:

LR: USS Forrest Royal

UL: USS Thuban UC: USS Constellation UR: USS James E. Keys

INDEX

—A—
Abbot; DD-629, 51
Abnaki; ATF-96, 22
Acadia; AD-42, 3
Adams, Charles F.; DDG-2, 68
Adams, John; SSBN-620, 136
Aeolus; ARC-3, 16
Aeolus; T-ARC-3, 162
Affray; MSO-511, 118
Ainsworth; FF-1090, 92
Ajax; AR-6, 15
Alabama; SSBN-731, 140, 141
Alamo; LSD-33, 108
Alamogordo; ARDM-2, 17
Alaska; SSBN-732, 141
Albany; CG-10, 27
Albany; SSN-753, 159
Albuquerque; SSN-706, 155
Alexandria; SSN-757, 159
Algol; LKA-54, 102
Allagash; AO-97, 10
Allen, Ethan; SSBN-608, 135
Altair; AKS-32, 9
Amber Jack; SS-522, 132
America; CV-66, 39, 40
America; CVA-66, 43, 44
American Legion; APA-17, 14
Amphion; AR-13, 16
Anchorage; LSD-36, 109
Anderson, Richard B.; DD-786, 56
Angler; SS-240, 121, 122
Annapolis; SSN-760, 160
Antelope; PG-86, 119
Antietam; CG-54, 30
Antietam; CVS-36, 49
Antrim; FFG-20, 96
Apopka; YTB-778, 163
Aquila; PHM-4, 120
Arcadia; AD-23, 2
Arcadia; AD-42, 3
Archerfish; AGSS-311, 124
Archerfish; SSN-678, 150
Arco; ARDM-5, 17
Arcturus; AF-52, 5
Argonaut; SS-475, 131
Aries; PHM-5, 120
Arkansas; CGN-41, 33
Arlington; AGMR-2, 8
Asheville; SSN-758, 160
Ashland; LSD-1, 107
Ashtabula; AO-51, 10
Aspro; SSN-648, 147
Assurance; AG-521, 7
Atlanta; SSN-712, 156
Atule; SS-403, 129
Augusta; SSN-710, 155
Ault; DD-698, 53
Austin; LPD-4, 104
Avalon; DSRV-2, 86
Avenger; MCM-1, 114
Aylwin; FF-1081, 91

—B—
Badger; DE-1071, 79
Badger; FF-1071, 90
Badoeing Straight; CVE-116, 44
Bagley; DE-1069, 79

Bagley; FF-1069, 90
Bainbridge; DLGN-25, 86
Baltimore; SSN-704, 154
Bancroft, George; SSBN-643, 139
Bang; SS-385, 128
Banner; AGER-1, 8
Barb; SSN-596, 144
Barbel; SS-580, 134
Barbero; SSG-317, 124
Barbey; FF-1088, 92
Barbour County; LST-1195, 114
Barney; DDG-6, 68
Barnstable County; LST-1197, 114
Barry; DD-933, 62
Barton; DD-722, 54
Bates, William H.; SSN-680, 150
Batfish; SS-310, 124
Batfish; SSN-681, 150
Baton Rouge; SSN-689, 151
Bauer; DE-1025, 76
Bausell; DD-845, 59
Baya; AGSS-318, 124
Beacon; PG-99, 119
Bearss; DD-654, 52
Beary, Donald B.; DE-1085, 81
Beary, Donald B.; FF-1085, 92
Beaufort; ATS-2, 23
Becuna; SS-319, 124, 125
Belknap; CG-26, 28
Belknap; DLG-26, 85
Belleau Wood; LHA-3, 102
Benham; DD-796, 57
Bennington; CVS-20, 49
Bergall; SS-320, 125
Bergall; SSN-667, 148
Berkeley; DDG-15, 69
Berry, Fred T.; DD-858, 60
Berry, Fred T.; DDE-858, 67
Biddle; CG-34, 29
Biddle; DLG-34, 85
Bigelow; DD-942, 63
Billfish; SSN-676, 149
Biloxi; CL-80, 34
Birmingham; SSN-695, 152
Blackfin; SS-322, 125
Blakely; DE-1072, 79
Blakely; FF-1072, 90
Blandy; DD-943, 63
Blue; DD-744, 54
Blueback; SS-581, 134
Bluefish; SSN-675, 149
Bluegill; SSK-242, 122
Blue Ridge; LCC-19, 102
Bolivar, Simon; SSBN-641, 138
Bolster; ARS-38, 17
Bonefish; SS-582, 134
Bonefish; SSR-582, 134
Boone; FFG-28, 97
Boone, Daniel; SSBN-629, 137
Borie; DD-704, 53
Boston; CAG-1, 26
Boston; SSN-703, 154
Boulder; LST-1190, 113
Bowditch; T-AGS-21, 161
Bowen; DE-1079, 80
Bowen; FF-1079, 91
Bowfin; SS-287, 123

Boxer; CVA-21, 42
Bradley; FF-1041, 87
Bradley, Robert G.; FFG-49, 99
Bream; SS-243, 122
Bremerton; SSN-698, 153
Brewton; DE-1086, 81
Brewton; FF-1086, 92
Bridget; DE-1024, 76
Briscoe; DD-977, 65
Brister; DER-327, 83
Bristol County; LST-1198, 114
Bronstein; DE-1037, 76
Bronstein; FF-1037, 87
Brooke; DEG-1, 82
Brooke; FFG-1, 94
Brough; DE-148, 75
Brown, Jessie L.; DE-1089, 81
Brown, Jessie L.; FF-1089, 92
Brownson; DD-868, 61
Brumby; FF-1044, 87
Brunswick; ATS-3, 24
Bryce Canyon; AD-36, 2
Buchanan; DDG-14, 69
Buffalo; SSN-715, 157
Bullhead; SS-332, 125
Bunker Hill; CG-52, 30
Burke, Arleigh; DDG-51, 73
Bushnell; AS-15, 19
Butte; AE-27, 4
Butternut; AN-9, 9
Byrd, Richard E.; DDG-23, 70

—C—
Cable, Frank; AS-40, 20
Cabot; CVL-28, 45
Cacapon; AO-52, 10
Cadmus; AR-14, 16
Calhoun, John C.; SSBN-630, 137
California; CGN-36, 32
Callaghan; DDG-994, 73
Caloosahatchie; AO-98, 11
Cambria; PA-36, 118
Camden; AOE-2, 13
Canisteo; AO-99, 11
Canopus; AS-34, 20
Cape Cod; AD-43, 3
Capodanno; FF-1093, 93
Carbonero; SS-337, 125
Caron; DD-970, 64
Carpenter; DD-825, 58
Carp; SS-338, 126
Carr; FFG-52, 99
Carver, George Washington; SSBN-656, 139
Cascade; AD-16, 1
Catawba; ATA-210, 22
Catfish; SS-339, 126
Catskill; MCS-1, 115
Cavalier; APA-37, 14
Cavalla; SS-244, 122
Cavalla; SSN-684, 150
Cayuga; LST-1186, 113
Cecil, Charles P.; DD-835, 58
Cecil, Charles P.; DDR-835, 74
Chancellorsville; CG-62, 32
Chandler, T.E.; DD-717, 54
Chandler; DDG-996, 73

Chara; AE-31, 5
Charleston; LKA-113, 103
Chauncey; DD-667, 52
Chemung; AO-30, 10
Chevalier; DD-451, 50
Chevalier; DD-805, 57
Chicago; CG-11, 27
Chicago; SSN-721, 157, 158
Chikaskia; AO-54, 10
Chosin; CG-65, 32
Chowanoc; ATF-100, 22
Cimarron; AO-22, 10
Cimarron; AO-177, 12
Cincinnati; SSN-693, 152
City of Corpus Christi; SSN-705, 154
Clamagore; SS-343, 126
Clark; FFG-11, 95
Clay, Henry; SSBN-625, 137
Cleveland; LPD-7, 104
Clymer, George; APA-27, 14
Coates; DE-685, 75
Cobbler; SS-344, 126
Cochrane; DDG-21, 70
Collett; DD-730, 54
Colhoun; DD-801, 57
Columbia; CL-56, 33
Columbus; CG-12, 27
Compass Island; AG-153, 7
Competent; AFDM-6, 6
Comstock; LSD-45, 110
Concord; AFS-5, 7
Cone; DD-866, 61
Conflict; MSO-426, 115
Conger; SS-477, 131
Connole; DE-1056, 77
Connole; FF-1056, 88
Conolly; DD-979, 66
Conquest; MSO-488, 118
Constant; MSO-427, 116
Constellation; Unknown, 164
Constellation; CV-64, 39, 50
Constellation; CVA-64, 43
Constitution; IX-21, 101
Conway; DDE-507, 67
Cony; DDE-508, 67
Conyngham; DDG-17, 69
Cook; APD-130, 15
Cook; DE-1083, 81
Cook; FF-1083, 92
Coontz; DDG-40, 72
Copeland; FFG-25, 97
Coral Sea; CV-43, 36
Coral Sea; CVA-43, 43
Coronado; AGF-11, 8
Coronado; LPD-11, 105
Corporal; SS-346, 126
Corpus Christi; SSN-705, 154
Corry; DD-817, 57
Corsair; SS-435, 130
Cowell; DD-547, 51
Cowpens; CG-63, 32
Criman; SS-323, 125
Croaker; SS-246, 122
Crockett; PG-88, 118
Crommelin; FFG-37, 98
Cubera; SS-347, 126

165

Curts; FFG-38, 98
Cushing; DD-985, 66
Cusk; SS-348, 126
Cutlass; SS-478, 131

—D—
Dace; SSN-607, 145
Dahlgren; DDG-43, 72, 73
Dale; CG-19, 27
Dallas; SSN-700, 153
Damato; DD-871, 61
Daniels, Josephus; CG-27, 28
Daniels, Josephus; DLG-27, 85
Darter; SS-576, 134
Dash; MSO-428, 116
David, Albert; DE-1050, 77
David, Albert; FF-1050, 88
Davidson; DE-1045, 76
Davidson; FF-1045, 87
Davis; DD-937, 62
Davis, Rodney M.; FFG-60, 101
De Grasse, Comte; DD-974, 65
De Haven; DD-727, 54
De Soto County; LST-1171, 111
De Wert; FFG-45, 99
Decator; DDG-31, 71
Denver; LPD-9, 104
Detector; MSO-429, 116
Detroit; AOE-4, 13
Devastator; MCM-6, 115
Dewey; DDG-45, 73
Dewey; DLG-14, 84
Deyo; DD-989, 67
Diablo; AGSS-479, 131
Diamond Head; AE-19, 4
Diodon; SS-349, 127
Direct; MSO-430, 116
Dixie; AD-14, 1
Dixon; AS-37, 20
Dogfish; SS-350, 127
Dolphin; SS-555, 133
Dominant; MSO-431, 116
Dortch; DD-670, 52
Downes; DE-1070, 79
Downes; FF-1070, 90
Doyle; FFG-39, 98
Drum; SSN-677, 149
Dubuque; LPD-8, 104
Duluth; LPD-6, 104
Duncan; FFG-10, 95
Du Pont; DD-941, 63
Durham; LKA-114, 103
Duval County; LST-758, 110
Dyess; DD-880, 61

—E—
Eastman, George; YAG-39, 163
Edenton; ATS-1, 23
Edison, Thomas A.; SSBN-610, 135
Edson; DD-946, 63
Edwards, Richard S.; DD-950, 63
Eisenhower, Dwight D., CVN-69, 46
Elk River; IX-501, 101
Elliot; DD-967, 64
Ellison, Harold J.; DD-864, 60
El Dorado; Unknown, 164
El Paso; LKA-117, 103
El Rod; FFG-55, 100
Engage; MSO-433, 116
England; CG-22, 28
England; DLG-22, 84

Enhance; MSO-437, 116
Entemedor; SS-340, 126
Enterprise; CVAN-65, 43
Enterprise; CVN-65, 45, 50
Enwald; LSD-6, 107
Epping Forest; LSD-4, 107
Epping Forest; MCS-7, 115
Erben; DD-631, 51
Escape; ARS-6, 17
Essex; CVA-9, 41
Essex; CVS-9, 48
Estocin; FFG-15, 96
Evans, Frank E.; DD-754, 55
Eversole; DD-789, 56
Excel; MSO-439, 117
Exultant; MSO-441, 117

—F—
Fahrion; FFG-22, 96
Fairfax County; LST-1193, 114
Falgout; DER-324, 83
Fanning; DE-1076, 80
Fanning; FF-1076, 91
Farragut; DDG-37, 72
Farragut; DLG-6, 83
Fearless; MSO-442, 117
Fidelity; MSO-443, 117
Fife; DD-991, 67
Finback; SSN-670, 149
Firedrake; AE-14, 3
Fiske; DD-842, 59
Fitch, Aubrey; FFG-34, 98
Flagstaff; PGH-1, 119
Flasher; SS-249, 122
Flasher; SSN-613, 145
Flatley; FFG-21, 96
Fletcher; DD-992, 67
Flint; AE-32, 5
Florida; SSBN-728, 140
Florikan; ASR-9, 21
Flying Fish; SSN-673, 149
Ford; FFG-54, 100
Forrestal; CV-59, 36, 37
Forrest Royal; Unknown, 163
Forster; DER-334, 83
Fortify; MSO-446, 117
Fort Fisher; LSD-40, 109
Fort Mandon; LSD-21, 107
Fort Marion; LSD-22, 107
Fort McHenry; LSD-43, 109
Fort Snelling; LSD-30, 108
Foster; Paul F., DD-964, 64
Fox; CG-33, 29
Fox; DLG-33, 85
Fox; Myles C., DDR-829, 74
Franklin; CV-13, 35
Franklin, Benjamin; SSBN-640, 138
Frederick; LST-1184, 113
Fresno; LST-1182, 112
Frigate Bird; MSC-191, 115
Fulton; AS-11, 18
Furer, Julius A.; FF-6, 86
Furer, Julius A.; FFG-6, 94
Furse; DD-882, 62

—G—
Gainard; DD-706, 53
Gallant; MSO-489, 118
Gallery; FFG-26, 97
Gallup; PG-85, 118
Galveston; CLG-3, 34
Gansevoort; DD-608, 51

Garcia; FF-1040, 87
Garrett County; LST-786, 110
Gary; FFG-51, 99
Gates, Thomas S.; CG-51, 30
Gato; SSN-615, 146
Gemini; PHM-6, 120
Georgia; SSBN-729, 140
Germantown; LSD-42, 109
Gettysburg; CG-64, 32
Gilbert Islands; CVE-107, 44
Gilmore, Howard W.; AS-16, 19
Glennon; DD-840, 59
Glover; AGDE-1, 7
Glover; AGFF-1, 8
Glover; FF-1098, 93
Goldsborough; DDG-20, 70
Gompers, Samuel; AD-37, 2
Goodrich; DDR-831, 74
Graham County; AGP-1176, 8
Grand Canyon; AR-28, 16
Grand Rapids; PG-98, 119
Grant County; LST-1174, 112
Grant, Ulysses S.; SSBN-631, 138
Grapple; ARS-53, 18
Grasp; ARS-51, 18
Gray; DE-1054, 77
Gray; FF-1054, 88
Grayback; LPSS-574, 134
Grayling; SSN-646, 147
Great Sitkin; AE-17, 4
Green Bay; PG-101, 119
Green, Nathanial; SSBN-636, 138
Greene, Eugene A.; DDR-711, 74
Greenlet; ASR-10, 21
Greenling; SS-213, 121
Greenling; SSN-614, 146
Greenwood; DE-679, 75
Gregory; DD-802, 57
Gridley; CG-21, 28
Groton; SSN-694, 152
Groves, Stephen W.; FFG-29, 97
Growler; SSG-577, 134
Guadalcanal; LPH-7, 106
Guadalupe; A0-32, 10
Guam; LPH-9, 106
Guardfish; SSN-612, 145
Gudgeon; SS-507, 132
Gudgeon; SS-567, 133
Guitarro; SSN-665, 148
Gurke; DD-783, 56
Gurnard; SSN-662, 148

—H—
Haddo; SSN-604, 145
Haddock; SSN-621, 146
Hale; DD-642, 52
Hale, Nathan; SSBN-623, 137
Haleakala; AE-25, 4
Halfbeak; SS-352, 127
Halibut; SSGN-587, 142
Halibut; SSN-587, 143
Hall, Gunston; LSD-5, 107
Hall, Gunston; LSD-44, 109
Hall, John L.; FFG-32, 97
Halsey; CG-23, 28
Halyburton; FFG-40, 98
Hamilton, Alexander; SSBN-617, 136
Hammerhead; SSN-663, 148
Hammond, Francis; DE-1067, 78
Hammond, Francis; FF-1067, 90
Hamner; DD-718, 54

Hamul; AD-20, 2
Hancock; CVA-19, 41, 42
Hancock, John; DD-981, 66
Hank; DD-702, 53
Hansford; APA-106, 14
Harder; SS-568, 133
Hardhead; SS-365, 127
Harkness; T-AGS-32, 161
Harlan County; LST-1196, 114
Hart, Thomas C.; DE-1092, 82
Hart, Thomas C.; FF-1092, 93
Hassayampa; AO-145, 11
Haverfield; DER-393, 83
Hawes; FFG-53, 99
Hawkbill; SSN-666, 148
Hawkins; DD-873, 61
Hayes; T-AGOR-16, 161
Hayler; DD-997, 67
Hector; AR-7, 15
Helena; SSN-725, 158
Henderson; DD-785, 56
Henley; DD-762, 55
Henry, Patrick; SSBN-599, 135
Hepburn; FF-1055, 88
Hercules; PHM-2, 120
Hermitage; LSD-34, 108
Hewes, Joseph; DE-1078, 80
Hewes, Joseph; FF-1078, 91
Hewitt; DD-966, 64
Higbee; DD-806, 57
High Point; PCH-1, 118
Hill, Harry W.; DD-986, 66
Hitchiti; ATF-103, 22
Hoel; DDG-13, 69
Hoist; ARS-40, 18
Holder; DD-819, 57
Holland; AS-32, 19, 20
Hollidaysburg; PCS-1385, 118
Hollister; DD-788, 56
Holt, Harold E.; DE-1074, 79
Holt, Harold E.; FF-1074, 90
Honolulu; SSN-718, 157
Hooper; DE-1026, 76
Horne; CG-30, 28
Horne; DLG-30, 85
Hornet; CVS-12, 48
Houston; SSN-713, 156
Houston, Sam; SSBN-609, 135
Hubbard, Harry E.; DD-748, 55
Hull; DD-945, 63
Hunley; AS-31, 19
Hunterdon County; LST-838, 110
Huntington, R.K.; DD-781, 56
Huse; DE-145, 75

—I—
Icefish; SS-367, 127
Illusive; MSO-448, 117
Implicit; MSO-455, 117
Inchon; LPH-12, 106
Independence; CV-62, 38
Indianapolis; SSN-697, 153
Inflict; MSO-456, 117
Ingersoll; DD-652, 52
Ingersoll; DD-990, 67
Ingraham; DD-694, 53
Ingraham; FFG-61, 101
Ingram, Jonas; DD-938, 62
Intercepter; YAGR-8, 163
Intrepid; CVS-11, 48
Investigator; YAGR, 163
Iowa; BB-61, 24

Irex; SS-482, 132
Isbel, Arnold J.; DD-869, 61
Isle Royal; AD-29, 2
Iwo Jima; LPH-2, 105

—J—
Jack; SSN-605, 145
Jackson, Andrew; SSBN-619, 136
Jackson, Henry M.; SSBN-730, 140
Jackson, Stonewall; SSBN-634, 138
Jacksonville; SSN-699, 153
James, Reuben; FFG-57, 100, 101
Jarrett; FFG-33, 97
Jason; AR-8, 15
Jefferson City; SSN-759, 160
Jefferson, Thomas; SSBN-618, 136
Johnston; DD-821, 58
Jones, John Paul; DDG-32, 71
Jouett; CG-29, 28
Jouett; DLG-29, 85
Joy, Turner; DD-951, 64
Juneau; LPD-10, 105
Jupiter; T-AKR-11, 162

—K—
Kaiser, Henry J.; T-AO-187, 162
Kalamazoo; AOR-6, 13
Kamehameha; SSBN-642, 139
Kane, Elisha Kent; T-AGS-27, 161
Kansas City; AOR-3, 13
Kauffman; FFG-59, 101
Kawishiwi; AO-146, 11
Kearsarge; CVS-33, 49
Kemper County; LST-854, 110
Kennedy, John F.; CV-67, 40, 41
Kennedy, John F.; CVA-67, 44
Kentucky; SSBN-737, 142
Keppler; DD-765, 55
Key, Francis Scott; SSBN-657, 139
Key West; SSN-722, 158
Keys, James E.; Unknown, 164
Kidd; DDG-993, 73
Kilauea; AE-26, 4
King; DDG-41, 72
King; DLG-10, 84
King, John; DDG-3, 68
Kingsport; T-AG-164, 160
Kinkaid; DD-965, 64
Kiowa; ATF-72, 22
Kirk; DE-1087, 81
Kirk; FF-1087, 92
Kiska; AE-35, 5
Kittyhawk; CV-63, 39
Kittywake; ASR-13, 21
Klakring; FFG-42, 98
Knox; DE-1052, 77
Knox; FF-1052, 88
Koelsh; DE-1049, 77
Koelsh; FF-1049, 88
Kraus, R.E.; DD-849, 60
Kretchmer; DER-329, 83
Kyes, James E.; DD-787, 56
Kyne; DE-744, 75

—L—
Lafayette; SSBN-616, 136
Laffey; unknown, 164
La Jolla; SSN-701, 153
La Moure County; LST-1194, 114
La Salle; AGF-3, 8
Lake Champlain; CG-57, 31
Lake Champlain; CVS-39, 49

Lake, Simon; AS-33, 20
Lamprey; SS-372, 127
Land, Emory S.; AS-39, 20
Lang; DE-1060, 78
Lang; FF-1060, 89
Lapon; SSN-661, 148
Lawe, William C.; DD-763, 55
Lawrence; DDG-4, 68
Leahy; CG-16, 27
Leary; DDR-879, 74
Lee, Robert E.; SSBN-601, 135
Lee, S.P.; T-AG-192, 160
Leftwich; DD-984, 66
Lester; DE-1022, 75
Lewis and Clark; SSBN-644, 139
Lexington; AVT-16, 24
Lexington; CVS-16, 49
Lexington; CVT-16, 50
Leyte; CVE-32, 44
Leyte Gulf; CG-55, 30, 31
Liberty; AGTR-5, 9
Liggett, Hunter; APA-14, 14
Lincoln, Abraham; CVN-72, 48
Lincoln, Abraham; SSBN-602, 135
Ling; SS-297, 123
Lipscomb, Glenard P.; SSN-685, 150
Litchfield County; LST-901, 110
Little Rock; CG-4, 26
Little Rock; CLG-4, 34
Lockwood; DE-1064, 78
Lockwood; FF-1064, 89
Lofberg; DD-759, 55
Long Beach; CGN-9, 32
Lorain County; LST-1177, 112
Los Alamos; AFDB-7, 6
Los Angeles; CA-135, 26
Los Angeles; SSN-688, 151
Louisville; SSN-724, 158
Luce; DDG-38, 72
Luce; DLG-7, 83
Luiseno; ATF-156, 23
Luzerne County; LST-902, 110

—M—
Maccaponi; AO-41, 10
MacDonough; DDG-39, 72
MacDonough; DLG-8, 84
Mackenzie, George K.; DD-836, 59
Mackeral; SS-570, 134
Madison, James; SSBN-627, 137
Magoffin; APA-199, 14
Mahan; DDG-42, 72
Mahan; DLG-11, 84
Manchester; CL-83, 34
Manitowoc; LST-1180, 112
Manley; DD-940, 63
Marathon; PG-89, 119
Marias; T-AO-57, 162
Marion, Francis; LPA-249, 103
Markab; AR-23, 16
Mars; AFS-1, 6
Marsh; DE-699, 75
Marshall, George C.; SSBN-654, 139
Marshall, John; SSBN-611, 136
Marshfield; T-AK-282, 162
Mataco; ATF-86, 22
Mathews; AKA-96, 9
Mauna Kea; AE-22, 4
McCaffery; DD-860, 60
McCain, John S.; DDG-36, 71

McCandless; FF-1084, 92
McCard, R.H.; DD-822, 58
McCloy; FF-1038, 87
McClusky; FFG-41, 98
McCormick, Lynde; DDG-8, 68
McDonnell, Edward; DE-1043, 76
McDonnell, Edward; FF-1043, 87
McInerney; FFG-8, 94
McKean; DD-784, 56
McKee; AS-41, 20
McMorris; DE-1036, 76
Memphis; SSN-691, 152
Mercury; T-AKR-10, 162
Mercy; T-AH-19, 161
Meredith; DD-890, 62
Merrill; DD-976, 65
Merrimack; AO-179, 12
Meyerkord; FF-1058, 89
Miami; SSN-755, 159
Michigan; SSBN-727, 140
Midway; CV-41, 35, 36
Miller; DE-1091, 80
Miller; FF-1091, 93
Miller, William C.; DE-259, 75
Milwaukee; AOR-2, 13
Minneapolis-Saint Paul; SSN-708, 155
Mirfak; T-AK-271, 161
Mispillion; AO-105, 11
Mississinewa; AO-144, 11
Mississippi; CGN-40, 33
Missouri; BB-63, 25
Mitscher; DDG-35, 71
Mitscher; DL-2, 83
Mobile; LKA-115, 103
Mobile Bay; CG-53, 30
Moctobi; ATF-105, 23
Moinester; DE-1097, 82
Moinester; FF-1097, 93
Molala; ATF-106, 23
Monob; YAG-61, 163
Monongahela; AO-178, 12
Monroe James; SSBN-622, 137
Monterey; CG-61, 31
Montgomery, Elmer; DE-1082, 81
Montgomery, Elmer; FF-1082, 91
Monticello; LSD-35, 108
Moore, John A.; FFG-19, 96
Moore, Samuel N.; DD-747, 54
Moosbrugger; DD-980, 66
Morison, Samuel Eliot; FFG-13, 95
Morton; DD-948, 63
Mount Baker; AE-34, 5
Mount Hood; AE-29, 5
Mount Katmai; AE-16, 3
Mount Vernon; LSD-39, 109
Mount Whitney; LCC-20, 102
Mullinix; DD-944, 63
Mystic; DSRV-1, 86

—N—
Narwhal; SSN-671, 149
Nashville; LPD-13, 105
Nassau; LHA-4, 102
Nautilus; SSN-571, 142
Navarro; APA-215, 14
Neches; AO-47, 10
Neosho; AO-143, 11
Neptune; ARC-2, 16
Neptune, SSN-525, 142
Neptune, T-ARC-2, 162
Nereus; AS-17, 19

Nevada; SSBN-733, 141
New; DD-818, 57
New Jersey; BB-62, 24, 25
New Orleans; LPH-11, 106
New York City; SSN-696, 152
Newport; LST-1179, 112
Newport News; CA-148, 26
Newport News; SSN-750, 158
Niagra Falls; AFS-3, 7
Nicholas; FFG-47, 99
Nicholson; DD-982, 66
Nimitz; CVAN-68, 44
Nimitz; CVN-68, 45
Nipmuc; ATF-157, 23
Nitro; AE-23, 4
Noa; DD-841, 59
Noble; APA-218, 15
Norfolk; SSN-714, 156
Normandy; CG-60, 31
Northampton; CC-1, 26
North Carolina; BB-55, 24
Norton Sound; AVM-1, 24
Norwalk; T-AK-279, 162
Nuclear Research Submarine; NR-1, 120

—O—
O'Bannon; DD-987, 66
O'Brien; DD-975, 65
O'Callahan; DE-1051, 77
O'Callahan; FF-1051, 88
Oak Hill; LSD-7, 107
Oak Ridge; ARDM-1, 17
Oberon; AKA-14, 9
Observation Island; EAG-154, 86
Odax; SS-484, 132
Ogden; LPD-5, 104
Ohio; SSBN-726, 140
Okinawa; LPH-3, 105
Oklahoma City; CG-5, 27
Oklahoma City; SSN-723, 158
Oldendorf; DD-972, 65
Olympia; SSN-717, 157
Omaha; SSN-692, 152
Opportune; ARS-41, 18
Orca; AVP-49, 24
Orion; AS-18, 19
Oriskany; CV-34, 35
Oriskany; CVA-34, 42
Orleck; DD-886, 62
Ortolan; ASR-22, 22
Ouellet; FF-1077, 91
Owens, Robert A.; DD-827, 58
Oxford; AGTR-1, 9
Ozark; MCS-2, 115
Ozbourn; DD-846, 59

—P—
Page County; LST-1076, 110
Page, Richard L.; DEG-5, 82
Page, Richard L.; FFG-5, 94
Palau; CVE-122, 44
Papago; ATF-160, 23
Parche; SS-384, 127
Parche; SSN-683, 150
Pargo; SSN-650, 147
Parks, Floyd B.; DD-884, 62
Parsons; DDG-33, 71
Pasadena; SSN-752, 159
Patterson; DE-1061, 78
Patterson; FF-1061, 89
Paul; DE-1080, 80

167

Paul; FF-1080, 91
Peary, Robert E.; DE-1073, 79
Peary, Robert E.; FF-1073, 90
Pegasus; PHM-1, 119
Peleliu; LHA-5, 102
Pennsylvania; SSBN-735, 142
Pensacola; LSD-38, 109
Peoria; LST-1183, 113
Perch; APSS-313, 124
Permit; SS-178, 120
Permit; SSN-594, 144
Perry, John R.; DE-1034, 76
Perry, N.K.; DD-883, 62
Perry, Oliver Hazard; FFG-7, 94
Peterson; DD-969, 64
Petrel; ASR-14, 21
Pharris; FF-1094, 93
Phelps; DD-360, 50
Philadelphia; SSN-690, 151
Philip, George; FFG-12, 95
Philippine Sea; CG-58, 31
Philippine Sea; CV-47, 36
Phoenix; SSN-702, 153
Pickaway; APA-222, 15
Picking; DD-685, 52
Piedmont; AD-17, 1
Pigeon; ASR-21, 22
Pintado; SSN-672, 149
Pirana; SS-389, 128
Pittsburgh; SSN-720, 157
Plainview; AGEH-1, 8
Platte; AO-186, 12
Pledge; MSO-492, 118
Pluck; MSO-464, 117
Plunger; SSN-595, 144
Plymouth Rock; LSD-29, 108
Pogy; SSN-647, 147
Point Defiance; LSD-31, 108
Point Loma; AGDS-2, 7
Polk, James K.; SSBN-645, 139
Pollack; SS-180, 120
Pollack; SSN-603, 145
Pomfret; SS-391, 128
Ponce; LPD-15, 105
Ponchatoula; AO-148, 12
Porpoise; SS-172, 120
Portland; LSD-37, 109
Portsmouth; SSN-707, 155
Power; DD-839, 59
Prairie; AD-15, 1
Pratt, William V.; DDG-44, 73
Pratt, William V.; DLG-13, 84
Preble; DDG-46, 73
Preble; DLG-15, 84
Preserver; ARS-8, 17
Prichett; DD-561, 51
Princeton; CG-59, 31
Princeton; CVS-37, 49
Proteus; AS-19, 19
Providence; SSN-719, 157
Puffer; SSN-652, 147
Puget Sound; AD-38, 2
Pulaski, Cassimir; SSBN-633, 138
Puller, Lewis B.; FFG-23, 96
Pyro; AE-24, 4

—Q—
Queenfish; SSN-651, 147

—R—
Racine; LST-1191, 114
Radford, Arthur W.; DD-968, 64

Raleigh; CL-7, 33
Raleigh; LPD-1, 104
Ramsey; DEG-2, 82
Ramsey; FFG-2, 94
Randolph; CVA-15, 41
Randolph; CVS-15, 49
Ranger; CV-61, 37, 38
Ranger; CVA-61, 43
Rankin; AKA-103, 9
Rasher; SS-269, 122
Rathburne; FF-1057, 88
Ray; SSN-653, 147
Ray; SSR-271, 123
Ray, David R.; DD-971, 65
Rayburn, Sam; SSBN-635, 138
Reasoner; DE-1063, 78
Reasoner; FF-1063, 89
Reclaimer; ARS-42, 18
Redfish; SS-395, 128
Redstone; T-AGM-20, 161
Reeves; CG-24, 28
Reeves; DLG-24, 84
Reid; FFG-30, 97
Remey; DD-688, 52
Remora; SS-487, 132
Rentz; FFG-46, 99
Requin; SS-481, 131
Requin; SSR-481, 132
Resolute; AFDM-10, 6
Revere, Paul; LPA-248, 103
Rich; DD-820, 58
Richard, Bon Homme; CVA-31, 42
Ricketts, Claude V.; DDG-5, 68
Rickover, Hyman G.; SSN-709, 155
Rigel; AF-58, 6
Rigel; T-AF-58, 160
Rivers, L. Mendel; SSN-686, 151
Roan, Charles H.; DD-853, 60
Roanoke; AOR-7, 14
Roanoke; CL-145, 34
Roark; DE-1053, 77
Roark; FF-1053, 88
Roberts, Samuel B.; DD-823, 58
Roberts, Samuel B.; FFG-58, 101
Robison; DDG-12, 69
Rodgers, John; DD-983, 66
Rogers; DD-876, 61
Rogers, Will; SSBN-659, 140
Ronquil; SS-396, 128
Roosevelt, Franklin D.; CVA-42, 43
Roosevelt, Theodore; CVN-71, 47, 48
Roosevelt, Theodore; SSBN-600, 135
Ross; DD-563, 51
Rowan; DD-405, 50
Runner; SS-476, 131
Rupertus; DD-851, 60
Rush, William R.; DD-714, 53
Rush, W.R.; DDR-714, 74
Russell, Richard B.; SSN-687, 151

—S—
Sabalo; SS-302, 123
Sacramento; AOE-1, 12
Safeguard; ARS-25, 17
Safeguard; ARS-50, 18
Saginaw; LST-1188, 113
Sailfish; SS-572, 133
Saint Louis; LKA-116, 103
Saint Paul; CA-73, 26

Saipan; LHA-2, 102
Salmon; SS-573, 133
Salt Lake City; SSN-716, 157
Salute; MSO-470, 118
Salvor; ARS-52, 18
Sample; DE-1048, 77
Sample; FF-1048, 87
Sampson; DDG-10, 68, 69
San Bernardino; LST-1189, 113
San Diego; AFS-6, 7
San Francisco; SSN-711, 155
San Jacinto; CG-56, 31
San Jacinto; CVL-30, 45
San Juan; SSN-751, 158
San Jose; AFS-7, 7
San Marcos; LSD-25, 107
San Onofre; ARD-30, 16
Sanctuary; AH-17, 9
Sand Lance; SS-381, 127
Sand Lance; SSN-660, 147
Santa Barbara; AE-28, 5
Saratoga; CV-3, 34
Saratoga; CV-60, 37
Saratoga; CVA-60, 43
Sarda; SS-488, 132
Sargo; SSN-583, 143
Savage; DER-386, 83
Savannah; AOR-4, 13
Sawfish; SS-276, 123
Scamp; SSN-588, 143
Schenectedy; LST-1185, 113
Schofield; DEG-3, 82
Schofield; FFG-3, 94
Scorpion, SSN-589, 143
Scott; DDG-995, 73
Sculpin; SSN-590, 144
Sea Cat; SS-399, 129
Sea Cliff; DSV-4, 86
Sea Devil; SS-400, 129
Sea Devil; SSN-664, 148
Sea Dog; SS-401, 129
Seadragon; SSN-584, 143
Sea Fox; SS-402, 129
Sea Horse; SSN-669, 148
Sea Owl; SS-405, 129
Sea Robin; SS-407, 129
Sea Wolf; SSN-575, 142
Seattle; AOE-3, 13
Segundo; SS-398, 129
Sellers; DDG-11, 69
Semmes; DDG-18, 69
Sentry; MCM-3, 115
Shadwell; LSD-15, 107
Shangri-La; CVA-38, 42
Shangri-La; CVS-38, 49
Shark; SSN-591, 144
Shasta; AE-6, 3
Shasta; AE-33, 5
Sheepshead; Unknown, 164
Shelton; DD-790, 57
Shenandoah; AD-26, 2
Shenandoah; AD-44, 3
Sherman, Forrest; DD-931, 62
Shields, Marvin; DE-1066, 78
Shields, Marvin; FF-1066, 89, 90
Shippingport; ARDM-4, 17
Shreveport; LPD-12, 105
Sides; FFG-14, 96
Sierra; AD-18, 1
Silversides; SS-236, 121
Silversides; SSN-679, 150
Simpson; FFG-56, 100

Sims, W.S.; DE-1059, 78
Sims, W.S.; FF-1059, 89
Sirius; T-AFS-8, 160
Skate, SSN-578, 142
Skipjack; SSN-585, 143
Skylark; ASR-20, 21
Small, Ernest G.; DDR-838, 74
Snapper, SS-185, 121
Snohomish County; LST-1126, 110
Snook; SSN-592, 144
Somers; DDG-34, 71
South Carolina; CGN-37, 33
South Dakota; BB-57, 24
Southerland; DD-745, 54
Spadefish, SSN-668, 148
Spartanburg County; LST-1192, 114
Spear, L.Y.; AS-36, 20
Spearfish; SS-190, 121
Sperry; AS-12, 18
Sperry, Chas S.; DD-697, 53
Sphinx; ARL-24, 17
Spiegel Grove; LSD-32, 108
Spinax; SSR-489, 132
Spot; SS-413, 130
Sprague, Clifton; FFG-16, 96
Sprauce; DD-963, 64
Springfield; CLG-7, 34
Standley, William H.; CG-32, 29
Standley, William H.; DLG-32, 85
Stark; FFG-31, 97
Stark County; LST-1134, 111
Steadfast; AFDM-14, 6
Stein; DE-1065, 78
Stein; FF-1065, 89
Steinaker; DD-863, 60
Sterett; CG-31, 29
Sterlet; SS-392, 128
Steuben, Von; SSBN-632, 138
Stickleback; SS-415, 130
Stimson, Henry L.; SSBN-655, 139
Stingray; SS-186, 121
Stoddard; DD-566, 51
Stoddert, Benjamin; DDG-22, 70
Stone County; LST-1141, 111
Strauss, Joseph; DDG-16, 69
Stribling; DD-867, 61
Strong; DD-758, 55
Stump; DD-978, 65
Sturgeon; SS-187, 121
Sturgeon; SSN-637, 146
Suffolk County; LST-1173, 111
Sumner, Allen M.; DD-692, 52
Sumter; LST-1181, 112
Sunbird; ASR-15, 21
Sunfish; SSN-649, 147
Suribachi; AE-21, 3
Sustain; AFDM-7, 6
Swenson, Lyman K.; DD-729, 54
Swordfish; SSN-579, 143
Sylvania; AFS-2, 6

—T—
Tacoma; PG-92, 119
Takelma; ATF-113, 23
Talbot; DEG-4, 82
Talbot; FFG-4, 94
Tallahatchie County; AVB-2, 24
Tallahatchie County; LST-1154, 11
Tang; SS-563, 133
Tarawa; LHA-1, 102
Tattnall; DDG-19, 70
Taurus; PHM-3, 120

Taussig, Joseph K.; DE-1030, 76
Tautog; SSN-639, 146
Taylor; FFG-50, 99
Tecumseh; SSBN-628, 146
Telfair; APA-210, 14
Tennessee; SSBN-734, 141
Texas; CGN-39, 33
Thach; FFG-43, 98
The Sullivans; DD-537, 50
Thomas, Herbert J.; DD-833, 58
Thomas, Lloyd; DD-764, 55
Thomason, John W.; DD-760, 55
Thomaston; LSD-28, 108
Thorn; DD-988, 67
Thresher; SSN-593, 144
Thuban; Unknown, 164
Ticonderoga; CG-47, 29
Ticonderoga; CVS-14, 35
Tidewater; AD-31, 2
Tigrone; SS-419, 130
Tillamook; ATA-192, 22
Tinosa; SSN-606, 145
Tioga County; LST-1158, 111
Tirante; SS-420, 130
Tiru; SS-416, 130
Tisdale, Mahlon S.; FFG-27, 97
Toledo; CA-133, 26
Topeka; CLG-8, 34
Topeka; SSN-754, 159
Towers; DDG-9, 68
Trenton; LPD-14, 105
Trepang; SS-412, 129
Trepang; SSN-674, 149
Trieste II; DSV-1, 86
Trigger; SS-564, 133
Tripoli; LPH-10, 106
Trippe; DE-1075, 79
Trippe; FF-1075, 91

Triton; SSN-586, 143
Triton; SSRN-586, 143
Trout; SS-566, 133
Truckee; AO-147, 12
Truett; DE-1095, 82
Truett; FF-1095, 93
Trumpetfish; SS-425, 130
Trutta; SS-421, 130
Truxton; CG-35, 29
Truxton; CGN-35, 32
Truxton; DLGN-35, 86
Tucker, Henry W.; DD-875, 61
Tulare; LKA-112, 103
Tullibee; SSN-597, 145
Tunny; APSS-282, 123
Tunny; SS-282, 123
Tunny; SSN-682, 150
Turner, Richard K.; CG-20, 27
Turtle; DSV-3, 86
Tuscaloosa; LST-1187, 113
Tusk; SS-426, 130
Twining; DD-540, 51

—U—
Underwood; FFG-36, 98
Utina; ATF-163, 23

—V—
Valdez; DE-1096, 82
Valdez; FF-1096, 93
Vallejo, Mariano G.; SSBN-658, 139
Valley Forge; CG-50, 30
Valley Forge; LPH-8, 106
Vancouver; LPD-2, 104
Vandegrift; FFG-48, 99
Vandenberg, Gen. Hoyt S.; T-AGM-10, 161

Vanguard; T-AGM-19, 161
Vega; AF-59, 6
Vernon County; LST-1161, 111
Vesole; DDR-878, 74
Vesuvius; AE-15, 3
Vincennes; CG-49, 30
Vinson, Carl; CVN-70, 46, 47, 50
Vireo; MSC-205, 115
Virginia; CGN-38, 33
Virgo; AE-30, 5
Voge; DE-1047, 77
Voge; FF-1047, 87
Vogelgesang; DD-862, 60
Vreeland; DE-1068, 79
Vreeland; FF-1068, 90
Vulcan; AR-5, 15

—W—
Wabash; AOR-5, 13
Waccamaw; AO-109, 11
Waccamaw; T-AO-109, 162
Waddell; DDG-24, 70
Wadsworth; FFG-9, 94, 95
Wahoo; SS-565, 133
Wainwright; CG-28, 28
Wainwright; DLG-28, 85
Waldo County; LST-1163, 111
Waldron; DD-699, 53
Waller; DD-466, 50
Warrington; DD-843, 59
Washington, George; SSBN-598, 135
Wasp; CV-7, 35
Wasp; CVS-18, 49
Watchman; AGR-16, 9
Waterford; ARD-5, 16
Webster, Daniel; SSBN-626, 137
Weeden; DE-797, 75

Weiss; APD-135, 15
West Virginia; SSBN-736, 142
Whale; SSN-638, 146
Whidbey Island; LSD-41, 109
Whipple; DE-1062, 78
Whipple; FF-1062, 89
Whipporwill; MSC-207, 115
White Plains, AFS-4, 7
White Sands; ARD-20, 16
Wichita; AOR-1, 13
Willamette; AO-180, 12
Williams, Jack; FFG-24, 96
Wilson, Henry B.; DDG-7, 68
Wilson, Robert; DD-847, 60
Wilson, Woodrow; SSBN-624, 137
Wiltsie; DD-716, 53
Wisconsin; BB-64, 25
Wood County; LST-1178, 112
Wood, William M.; DDR-715, 74
Worden; CG-18, 27
Wren; DD-568, 51
Wright; CC-2, 26

—Y—
Yarnell, Harry E.; CG-17, 27
Yarnell, Harry E.; DLG-17, 84
Yellowstone; AD-41, 3
YFRT-520, 163
York County; LST-1175, 112
York Town; CG-48, 29
York Town; CV-5, 35
York Town; CVS-10, 48
Yosemite; AD-19, 1, 2
Young, John; DD-973, 65

—Z—
Zellars; DD-777, 56